STUDENT'S SOLUTION MANUAL

PRINCIPLES OF BIOCHEMISTRY

ZUBAY/PARSON/VANCE

Prepared by

Hugh Akers
Lamar University

Caroline Breitenberger
Ohio State University

WCB Wm. C. Brown Publishers

Dubuque, IA Bogota Boston Buenos Aires Caracas Chicago
Guilford, CT London Madrid Mexico City Sydney Toronto

A Times Mirror Company

ISBN 0-697-22870-3

Printed in the United States of America by Times Mirror Higher Education Group, Inc.,
2460 Kerper Boulevard, Dubuque, Iowa, 52001

TABLE OF CONTENTS

Part 7 Storage and Utilization of Genetic Information

1 Cells, Biomolecules and Water

Summary

In this chapter we discussed the ways in which biochemistry parallels ordinary chemistry and those in which it is quite different. The chief points to remember are the following:

1. The basic unit of life is the cell, which is a membrane-enclosed, microscopically visible object.
2. Cells are composed of small molecules, macromolecules, and organelles. The most prominent small molecule is water, which constitutes 70% of the cell by weight. Other small molecules are present only in quite small amounts; they are precursors or breakdown products of macromolecules or coenzymes. There are four types of macromolecules: lipids, carbohydrates, proteins, and nucleic acids.
3. Noncovalent intermolecular forces largely determine the folded structure adopted by a macromolecule, particularly the relative affinity of different groupings on the macromolecule for water. In general, hydrophobic groupings are buried within the folded macromolecular structure, whereas hydrophilic groupings are located on the surface, where they can interact with water.
4. Biochemical reactions utilize a limited number of elements, most prominently carbon, hydrogen, oxygen, nitrogen, sulfur, and phosphorus. Many biochemical reactions are simple organic reactions.
5. Biochemical reactions are carried out under very mild conditions in aqueous solvent. The reactions can proceed under these conditions because of the highly efficient nature of protein enzyme catalysts.
6. Biochemical reactions frequently require energy. The most common source of chemical energy used is adenosine triphosphate (ATP). The splitting of a phosphate from the ATP molecule can provide the energy needed to make an otherwise unfavorable reaction proceed in the desired direction.
7. Biochemical reactions of different types are localized to different parts in the cell.
8. Biochemical reactions are frequently organized into multistep pathways.
9. Biochemical reactions are regulated according to need by controlling the amount and activity of enzymes in the system.
10. Most organisms depend on other organisms for their survival. Frequently this is because a given organism cannot make all of the compounds needed for its growth and survival.
11. The specific properties of any protein are due to the specific sequence of amino acids in its polypeptide chains. This sequence is determined by the genetic information carried by the sequence of DNA nucleotides. DNA transfers the information to messenger RNA, which serves as the template for protein synthesis.
12. Shortly after the earth was formed and had cooled to a reasonable temperature, chemical processes produced compounds that would be used in the development of living cells. Nucleic acids are the most important compounds for living cells, and it is believed that they played the central role in the origin of life.
13. Evolutionary trees based on morphology or biochemical differences indicate that all living systems are related through a common evolution.

2 Thermodynamics in Biochemistry

Summary

In this chapter we discussed some principles of thermodynamics as they relate to biochemical reactions. The following points are of greatest importance.

1. Thermodynamics is useful in biochemistry for predicting whether a given reaction can occur and, if so, how much work a cell can obtain from the process.
2. The thermodynamic quantities energy, enthalpy, entropy, and free energy are properties of the state of a system. Changes in these quantities depend only on the difference between the initial and final states, not on the mechanism whereby the system goes from one state to the other.
3. Energy is the capacity to do work.
4. The first law of thermodynamics says that energy cannot be created or destroyed in a chemical reaction. If the energy of a system increases, the surroundings must lose an equivalent amount of energy, either by the transfer of heat or by the performance of work.
5. The energy of a molecule includes translational, rotational, and vibrational energy, as well as electronic and nuclear energy. Electronic terms usually account for most of the change in energy ΔE in a chemical reaction.
6. The change in enthalpy ΔH is given by the expression $\Delta H = \Delta E + \Delta(PV)$. For most biochemical reactions, ΔE and ΔH are nearly equal. The organic molecules found in cells generally have much higher enthalpies than the simpler molecules from which they are built.
7. In most reactions that proceed spontaneously, the enthalpy of the system decreases. If no work is done, the system gives off heat to the surroundings. But in some spontaneous reactions, heat is absorbed in the absence of work and the enthalpy of the system increases. Such reactions invariably show an increase in the entropy of the system.
8. Entropy is a measure of the order in a system: Systems that are highly ordered have low entropies. The entropy of a molecule depends mainly on translational and rotational freedom. Biological macromolecules generally have much lower entropies than their building blocks.
9. The second law of thermodynamics states that an overall increase must take place in the entropy of the system and its surroundings in any process that occurs spontaneously. An isolated system proceeds spontaneously to states of increasingly greater entropy (greater disorder).
10. The change in the free energy of a system is defined as $\Delta G = \Delta H - T\Delta S$, where T is the absolute temperature. A reaction at constant pressure and temperature can occur spontaneously if, and only if, ΔG is negative. The maximal amount of useful work that can be obtained from a reaction is equal to $-\Delta G$.
11. For a reaction in solution, ΔG depends on the standard free energy change (ΔG°) and on the concentrations of the reactants and products. The standard free energy change is related to the equilibrium constant by the expression $\Delta G^\circ = -RT \ln K_{eq}$. Increasing the concentration of the reactants relative to the concentration of the products makes ΔG more negative.
12. Reactions that are thermodynamically unfavorable can be coupled to favorable reactions. The coupling of the reactions may be direct or sequential, as in a biochemical pathway.
13. ATP is the main coupling agent for free energy in living cells. The free energy provided by the hydrolysis of ATP is used to drive many reactions that would not occur spontaneously by themselves.

14. Several features make ATP particularly well suited for its role. First, hydrolysis of ATP to ADP and P_i or to AMP and PP_i releases a considerable amount of free energy. Second, ATP does not hydrolyze rapidly by itself, but it can be hydrolyzed readily in enzymatically catalyzed reactions. This difference allows the free energy of hydrolysis to be channeled into reactions in which it is needed but to be conserved when energy is not in demand. Third, the products of the hydrolysis of ATP provide opportunities for coupling to a wide variety of chemical reactions. Finally, the adenine and ribosyl groups of ATP, ADP, and AMP provide additional structural features that allow these molecules to bind to a large number of enzymes and thus to participate in regulating enzymatic activities.

Problems

1. In some respects, thermodynamic considerations are more important to a biochemist than to a chemist. Why is this so?

2. Name three extensive and three intensive properties that relate to thermodynamic quantities. What is the basic difference between the two types of properties?

3. What is meant by a state function? Why is enthalpy a state function?

4. Why can we equate internal energy and enthalpy for most biochemical reactions?

5. In thermodynamics it is important to distinguish between the total system and the system being studied. Why?

6. Why do enthalpies and entropies of solvation tend to negate one another?

7. Why does the entropy of a weak acid decrease on ionization?

8. Transfer of a hydrophobic molecule (e.g., a hydrophobic amino acid and side chain) from an aqueous to a nonaqueous environment is entropically favorable. Explain.

9. As we will see in chapter 13, oxaloacetate is formed by the oxidation of malate.

$$\text{L-Malate} + NAD^+ \rightarrow \text{Oxaloacetate} + NADH + H^+$$

The reaction has a $\Delta G^{\circ\prime}$ of +7.0 kcal/mole. Suggest reasons that the reaction proceeds in the direction of oxaloacetate production in the cell.

10. Cite three factors that make ATP ideally suited to transfer energy in biochemical systems.

11. You wish to measure the $\Delta G^{\circ\prime}$ for hydrolysis of ATP,

$$ATP + H_2O \rightarrow ADP + P_i + H^+$$

but the equilibrium for the hydrolysis lies so far toward products that analysis of the ATP concentration at equilibrium is neither practical nor accurate. However, you have the following data that allow calculation of the value indirectly.

$$\text{Creatine phosphate} + ADP + H^+ \rightarrow ATP + \text{Creatine} \quad K'_{eq} = 59.5 \qquad \textbf{(P1)}$$

$$\text{Creatine} + P_i \rightarrow \text{Creatine phosphate} + H_2O \quad \Delta G^{\circ\prime} = +10.5 \text{ kcal/mole} \qquad \textbf{(P2)}$$

Assume that $2.3RT = 1.36$ kcal/mole

a. Calculate the value of $\Delta G^{\circ\prime}$ for reaction (P1).
b. Calculate the $\Delta G^{\circ\prime}$ for the hydrolysis of ATP.

12. The hydrolysis of lactose (D-galactosyl-β-(1,4) D-glucose) to D-galactose and D-glucose occurs with $\Delta G^{\circ\prime}$ of –4.0 kcal/mole.
    a. Calculate K'_{eq} for the hydrolytic reaction.
    b. What are the $\Delta G^{\circ\prime}$ and K'_{eq} for the synthesis of lactose from D-galactose and D-glucose?
    c. Lactose is synthesized in the cell from UDP-galactose plus D-glucose and is catalyzed by lactose synthase. Given that $\Delta G^{\circ\prime}$ of hydrolysis of UDP-galactose is –7.3 kcal/mole, calculate $\Delta G^{\circ\prime}$ and K'_{eq} for the reaction

$$\text{UDP-galactose} + \text{D-glucose} \rightarrow \text{Lactose} + \text{UDP}$$

13. For each of the following reactions, calculate $\Delta G^{\circ\prime}$ and indicate whether the reaction is thermodynamically favorable as written.
    a. Glycerate-1,3-bisphosphate + Creatine → Phosphocreatine + 3-Phosphoglycerate
    b. Glucose-6-phosphate → Glucose-1-phosphate
    c. Phosphoenolpyruvate + ADP → Pyruvate + ATP

14. Assume that an individual needs 2,500 Calories per day (1 Calorie = 1,000 calories or 1 kcal) to meet energy requirements. For simplicity, consider that the energy needs of this individual are met with glucose (not that unrealistic, considering that most of the world meets these needs mainly with starches). Glucose, and carbohydrates in general contain about 4 Calories per gram. The ATP yield during catabolism is 30 moles of ATP per mole of glucose. What mass (in pounds) of K_4ATP are synthesized per day by this individual?

15. Proteins that serve as gene repressors frequently have two identical binding sites that interact with complementary sites on the DNA. If a repressor protein is cut in half without damaging either of its DNA binding sites, how is the binding to DNA affected? How is the binding affected if one of the two binding sites on the DNA is eliminated by changing the nucleotide sequence? Discuss the enthalpy, entropy, and free energy effects.

Solutions

1. Chemists can often push a thermodynamically unfavorable reaction to proceed to some extent by manipulating experimental reaction parameters such as pressure, temperature, or concentrations of reactants. Biochemists generally cannot alter such reaction parameters because organisms function within limited concentration ranges and at essentially constant pressure and temperature.
3. A state function describes the thermodynamic parameters of the system under consideration at a particular moment. Only the difference between initial and final states, not the path taken to achieve these states, is important in most thermodynamic considerations. Enthalpic contributions defining the thermodynamic state are considered only at the initial and final states. Enthalpy is independent of pathway and is therefore a state function.
5. In thermodynamics, the total system is the universe and it consists of a particular system and its environment. If the particular system under study is a closed system, it can exchange internal energy as heat and work with its surroundings, but no exchange of matter can occur. An open system can also exchange matter with its surroundings. Therefore, although any energy lost by a particular system is gained by its environment (and vice versa), the energy of the universe remains constant. This is the first law of thermodynamics. It is important to distinguish between the system and the universe in order to differentiate between a situation in which energy can change and a situation in which energy is constant.

   Another aspect to be considered is the entropy. For a system, the entropy can increase, decrease, or stay the same. Consequently, the entropy of the environment can decrease, increase, or stay the same. However, the entropy of the universe is always increasing to a maximum. This is the second law of thermodynamics.

   Living organisms are open systems and can exchange heat, work, and matter with their environment. They are highly ordered (low in entropy) and maintain this order at the expense of an increase in entropy in the environment.
7. Entropy decreases because of hydration effects. The ionized species will order much of the water during hydration and result in a decrease in the total number of free molecules.
9. The reaction shown describes the formation of oxaloacetate from the oxidation of malate in the TCA or Krebs cycle. Given the large (+) $\Delta G^{\circ\prime}$ value for the reaction, the reaction lies strongly in favor of the reactants at standard state equilibrium. However, conditions in the cell impose a steady state whose conditions are far removed from the thermodynamic standard state. The reaction will proceed toward oxaloacetate formation in the cell if the concentration of products is kept small. This is accomplished in the mitochondria in two ways.
   a. NADH is oxidized by the mitochondrial electron transport system, thereby diminishing the product (NADH) concentration and replenishing the reactant (NAD^+) concentration.

b. Oxaloacetate is condensed with acetyl-CoA in the citrate synthase-catalyzed formation of citrate. This condensation is thermodynamically favorable (approximately –8 kcal/mole). Coupling citrate synthase and malate dehydrogenase through the common intermediate oxaloacetate is an example of an exergonic (thermodynamically favorable) reaction driving an endergonic (thermodynamically unfavorable) reaction.

11. a. The standard state free energy change can be calculated from the equilibrium constant,

$$\Delta G^{\circ\prime} = -2.3\ RT \log K'_{eq}$$

$$\Delta G^{\circ\prime} = -\ 1.36\ \text{kcal/mole}\ (\log 59.5)$$

$$\Delta G^{\circ\prime} = -2.4\ \text{kcal/mole}$$

b. The hydrolysis of ATP can be determined by the combination of reactions (P3) and (P4) as follows:

(1) Creatine phosphate + ADP + $H^+ \rightarrow$ ATP + creatine + $\Delta G^{\circ\prime} = -2.4$ kcal/mole **(P3)**
(2) Creatine + $P_i \rightarrow$ creatine phosphate + H_2O $\Delta G^{\circ\prime} = +10.5$ kcal/mole **(P4)**
(3) H^+ + ADP + $P_i \rightarrow$ ATP + H_2O $\Delta G^{\circ\prime} = +7.9$ kcal/mole **(P5)**

Therefore, hydrolysis is the reverse of reaction (P5).

$$\text{ATP} + H_2O \rightarrow \text{ADP} + P_i + H^+\ \Delta G^{\circ\prime} = -7.9\ \text{kcal/mole}$$

13. a. Glycerate-1,3-bisphosphate $H_2O \rightarrow$ glycerate-3-phosphate + P_i ($\Delta G^{\circ\prime} = -11.8$ kcal/mole)

Creatine + $P_i \rightarrow$ phosphocreatine + H_2O $\Delta G^{\circ\prime} = +10.3$ kcal/mole

Glycerate-1,3-bisphosphate + creatine $\rightarrow$ phosphocreatine + glycerate-3-phosphate $\Delta G^{\circ\prime}$ –1.5 kcal/mole

Thermodynamically favorable as written.

b. D-glucose-6-P_i + $H_2O \rightarrow$ D-glucose + P_i $\Delta G^{\circ\prime} = -3.3$ kcal/mole
D-glucose + $P_i \rightarrow$ D=glucose-1-P_i +H_2O $\Delta G^{\circ\prime} = +5.0$ kcal/mole
D-glucose-6-$P_i \rightarrow$ D-glucose-1-P_i $\Delta G^{\circ\prime} = +1.7$ kcal/mole
Thermodynamically favorable as written.
(The phosphate linkage to the 6-OH group of glucose is a phosphate ester (a phosphorylated alcohol). The phosphate linkage to the C-1 position is a phosphorylated hemiacetal OH, thus the difference in free energy of hydrolysis of the two compounds.)

c. Phospheonolpyruvate (PEP) + $H_2O \rightarrow$ pyruvate + P_i $\Delta G^{\circ\prime} = -14.8$ kcal/mole
H^+ + ADP + $P_i \rightarrow$ ATP + H_2O $\Delta G^{\circ\prime} = +7.5$ kcal/mole
H^+ + PEP + ATP $\rightarrow$ ATP + pyruvate $\Delta G^{\circ\prime} = -7.3$ kcal/mole
Thermodynamically favorable as written.

15. In the first case, where the repressor protein is cut in half, the binding enthalpy for each part would be essentially half the enthalpy value for the intact repressor. The entropy would be less favorable (less positive) because of the chelation effect. As a result, the free energy would be less favorable (less negative).

In the second case, where one of the binding sites on the DNA is eliminated, the binding enthalpy for the repressor would be approximately half that for repressor binding to the unmodified DNA sequence. The entropy would depend on the extent of hydration and the extent of mobility of the unbound portion of the repressor. Again, one would expect the free energy for binding to be less favorable.

3 The Building Blocks of Proteins: Amino Acids, Peptides, and Polypeptides

Summary

In this chapter we dealt with some of the fundamental properties of amino acids and polypeptide chains. The following points are especially important.

1. Nineteen of the 20 amino acids commonly found in proteins have a carboxyl group and an amino group attached to an α-carbon atom; they differ in the side chain attached to the same α carbon.
2. All amino acids have acidic and basic properties. The ratio of base to acid form at any given pH can be calculated from the p*K* with the help of the Henderson-Hasselbach equation.
3. All amino acids except glycine are asymmetric and therefore can exist in at least two different stereoisomeric forms.
4. Peptides are formed from amino acids by the reaction of the α-amino group from one amino acid with the α-carboxyl group of another amino acid.
5. Polypeptide formation involves a repetition of the process involved in peptide synthesis.
6. The amino acid composition of proteins can be discovered by first breaking down the protein into its component amino acids and then separating the amino acids in the mixture for quantitative estimation.
7. The amino acid sequences of proteins can be discovered by breaking down the protein into polypeptide chains and then partially degrading the polypeptide chains. For each polypeptide chain fragment, the sequence is determined by stepwise removal of amino acids from the amino-terminal end of the polypeptide chain. Two different methods of forming polypeptide chain fragments are used so as to produce a map of overlapping fragments, from which the sequence of undegraded polypeptide chains in the proteins can be deduced.
8. Polypeptide chains with a predetermined amino acid sequence can be synthesized by chemical methods involving carboxyl-group activation.

Problems

1. A typical protein is 16% nitrogen by weight. How well does this percentage value match with the formula for proteins proposed by Berzelius?

2. a. A 10mM solution of a weak monocarboxylic acid has a pH of 3.00. Calculate the values for K_a and pK_a for this carboxylic acid.
   b. You add 0.06 g of NaOH (M_r = 40) to 1,000 ml of the acid solution in part (a). Calculate the final pH assuming no volume change.

3. A buffer was prepared by dissolving 3.71 g of citric acid and 2.91 g of KOH in water and diluting to a final volume of 250 ml. What is the pH of this buffer? What is the $[H^+]$? Use 3.14, 4.77, and 6.39 for the pK_a's of citric acid.

4. Given the pK_a values in the text, predict how the titration curves for glutamic acid and glutamine differ.

5. Calculate the isoelectric point for histidine, aspartic acid, and arginine. Calculate the fractional charge for each ionizable group on aspartate at pH equal to the pI. Do these calculations verify the isoelectric point of aspartic acid?

6. Which of the naturally occurring amino acid side chains are charged at pH 2? pH 7? pH 12? (Consider only those amino acids whose side chains have >10% charge at the pH examined.)

7. Amino acids are sometimes used as buffers. Indicate the appropriate pH value(s) of buffers containing aspartic acid, histidine, and serine.

8. Polyhistidine is insoluble in water at pH 7.8 but is soluble at pH 5.5. Explain this observation. Would you expect the polymer to be soluble at pH 10?

9. As indicated in the text, 19 of the 20 amino acids have a chiral α carbon with only glycine (R = H) lacking chirality. Two of the protein amino acids have a second chiral carbon. Can you identify them?

10. A mixture of alanine, glutamic acid, and arginine was chromatographed on a weakly basic ion-exchange column (positively charged) at pH 6.1. Predict the order of elution of the amino acids from the ion-exchange column. Are the amino acids separated from each other? Explain.
    Suppose you have a weakly acidic ion-exchange column (negatively charged), also at pH 6.1. Predict the order of elution of the amino acids from this column. Propose a strategy for separating the amino acids using one or both columns. Explain your rationale. (Assume only ionic interactions between the amino acids and the ion-exchange resin.)

11. You have a peptide that is a potent inhibitor of nerve conduction and you wish to obtain its primary sequence. Amino acid analysis reveals the composition to be Ala(5); Lys; Phe. Reaction of the intact peptide with FDNB releases free DNP-alanine on acid hydrolysis. ε-DNP-lysine (but not α-DNP-lysine) is also found. Tryptic digestion gives a tripeptide (composition Lys, Ala(2)) and a tetrapeptide (composition Ala(3), Phe). Chymotryptic digestion of the intact peptide releases a hexapeptide and free alanine. Derive the peptide sequence.

12. From a rare fungus you have isolated an octapeptide that prevents baldness, and you wish to determine the peptide sequence. The amino acid composition is Lys(2), Asp, Tyr, Phe, Gly, Ser, Ala. Reaction of the intact peptide with FDNB yields DNP-alanine plus 2 moles of ε-DNP-lysine on acid hydrolysis. Cleavage with trypsin yields peptides the compositions of which are (Lys, Ala, Ser) and (Gly, Phe, Lys), plus a dipeptide. Reaction with chymotrypsin releases free aspartic acid, a tetrapeptide with the composition (Lys, Ser, Phe, Ala), and a tripeptide the composition of which, following acid hydrolysis, is (Gly, Lys, Tyr). What is the sequence?

Solutions

1. Berzelius' proposal of $C_{40}H_{62}N_{10}O_{12}$ has a molecular mass of 874 g/mole of which 140 g/mole is nitrogen or:

$$\frac{(140 \text{ g N/mole protein})(100)}{874 \text{ g/mole protein}} = 16.0\ \%\ \text{nitrogen by mass}$$

The match is perfect.

3. $$\frac{3.71 \text{ g citric acid}}{192 \text{ g citric acid/mole}} = 0.0193 \text{ mole citric acid}$$

$$\frac{2.91 \text{ g KOH}}{56.1 \text{ g KOH/mole}} = 0.0519 \text{ mole KOH}$$

after reaction
0.0060 mole $HCit^{2-}$
0.0133 mole Cit^{3-}

Using the Henderson-Hasselbach equation:

$$pH = 6.39 + \log\frac{0.0133 \text{ mole/0.25 L}}{0.0060 \text{ mole/0.25 L}} = 6.39 + 0.35 = 6.74$$

$$[H^+] = 10^{-6.74} = 1.82 \times 10^{-7} \text{ M}$$

5. The amino acid has no net charge at the isoelectric pH. The pI value of amino acids with side chains that are not ionizable are isoelectric at pH that is the arithmetic mean of the p*K*s of the carboxyl and the α-amino groups. For example, the pI of glycine is (2.35 + 9.78)/2 = 6.07. Amino acids whose R-groups are ionizable may be divided between those whose side chains contribute to the positive charge and those contributing to the negative charge. If the R-group contributes positive charge (basic amino acid), the fractional charge residing on the α-amino group and on the R-group must total 1. This is achieved when the pH is the arithmetic mean of pK_{amino} plus pK_R. Similarly, the isoelectric pH of an acidic amino acid is equal to the arithmetic mean of $pK_{carboxyl}$ and pK_R.

$$\text{pI histidine} = [(pK_{amino}) + (pK_{imidazole})]/2$$
$$\text{pI} = 7.69$$

$$\text{pI aspartic acid} = [(pK_{carboxyl}) + (pK_R)]/2$$
$$\text{pI} = 2.95$$

$$\text{pI arginine} = (pK_{amino} + pK_R)/2$$
$$\text{pI} = 10.74$$

Apply the Henderson-Hasselbach equation to calculate the ratio of unprotonated to protonated groups on aspartate, then calculate the fractional charge on each group. At pH = pI (2.95), the α-amino group will be virtually fully protonated and will contribute one positive charge. The fractional charge on the carboxyl groups are

$$\alpha-\text{COOH}: \text{p}K_a = 1.99$$

$$\text{pH} = \text{p}K_a + \log\left[\frac{\text{A}^-}{\text{HA}}\right]$$

$$2.95 = 1.99 + \log\left[\frac{\text{A}^-}{\text{HA}}\right]$$

$$\log\left[\frac{\text{A}^-}{\text{HA}}\right] = 0.96$$

$$\left[\frac{\text{A}^-}{\text{HA}}\right] = 9.1, \text{ and } \left[\text{A}^-\right] + \left[\text{HA}\right] = 1$$

$[\text{A}^-] = 0.9$ fractional negative charge (90% of the α-carboxylate will be unprotonated at any time)

A similar calculation will reveal that the fractional negative charge on the *β*-carboxylate will be significantly less because the pH (2.95) is below the pK_a. Hence,

$[\text{A}^-] = 0.1$ negative charge on the *β*-carboxylate

Sum positive and negative charge contributions are

α-amino group	α-carboxyl	*β*-carboxyl
(+1)	(–0.9)	(–0.1) = 0

These data demonstrate that the net charge on aspartic acid is zero at pH 2.95 and thus verifies 2.95 as the isoelectric pH.

7. The buffering capacity of an ionizable group is greatest when the pH is equal to pK_a for the group and becomes inconsequential at pH values +/–1 pH unit from pK_a. In principle, we must consider the potential buffering capacity for each ionizable group.

   Aspartic acid: pH 2, 4, 10. (The pH range 1 to 5 will be buffered by the α-carboxyl and the *β*-carboxyl groups.)

   Histidine: pH 2, 9, 6 (imidazole side chain).

   Serine: pH 2, 9. (The pK_a of the alcohol is outside the range of pH normally considered for buffers.)

9. Threonine and isoleucine both contain two chiral carbons.

11. FDNB reacts with primary amines and is used to label the N-terminal amino acid. Were lysine the N-terminal, the diDNP derivative would have been observed. Alanine is the other amino acid that reacted with FDNB and is the N-terminal. Thus (Ala, Ala_4, Lys, Phe), where (–) indicates the peptide bond between two amino acids in sequence and (,) separates amino acids whose positions in the sequence have not been determined.

    Trypsin cleaves the peptide bonds in which lysine or arginine contribute the carboxyl group. The tripeptide derived from trypsin cleavage must have the sequence (Ala-Ala-Lys). The sequence of the tetrapeptide cannot be determined at this point.

Chymotrypsin cleaves most rapidly at the carboxyl side of Phe, Tyr, or Trp residues in peptides. The release of only Ala and an intact hexapeptide is consistent with the bonding (Phe-Ala) on the C-terminal of the peptide.

Based on the data, propose the following sequence and determine if the sequence is consistent with the data.

	Ala-Ala-Lys-Ala-Ala-Phe-Ala
FDNB/Hydrolysis	α-DNP-Ala; ε-DNP-Lys
Trypsin	(Ala-Ala-Lys) (Ala_3, Phe)
Chymotrypsin	(Ala-Ala-Lys-Ala-Ala-Phe) (Ala)

The proposed sequence is consistent with the data.

4 The Three-Dimensional Structures of Proteins

In this chapter we introduced some of the basic principles that govern protein structure. The discussion of protein structures begun in this chapter is continued in many other chapters in this text in which we consider structures designed for specific purposes. In chapter 5 we examine the protein structures for two systems: the protein that transports oxygen in the blood and the proteins that constitute muscle tissue. In chapters 8 and 9 we discuss structures of specific enzymes. In chapters 17 and 24 we consider proteins that interact with membranes. In chapters 30 and 31 we study regulatory proteins that interact with specific sites on the DNA. And finally, in supplement 3 we examine the structures of immunoglobin molecules.

In this chapter we also introduced the subject of the three-dimensional structures of proteins. This is an important subject to keep in mind throughout the text. Our discussion focused on the following points.

1. Most proteins may be divided into two groups: fibrous and globular. Fibrous proteins usually serve structural roles. Globular proteins function as enzymes and in many other capacities.
2. The three most prominent groups of fibrous proteins are the α-keratins, the β-keratins, and collagen.
3. The α-keratins are composed of right-handed helical polypeptide chains in which all the peptide NH and carbonyl groups form intramolecular hydrogen bonds. When these helical coils interact, they form left-handed coiled coils.
4. The β-keratins consist of extended polypeptide chains in which adjacent polypeptides are oriented in either a parallel or an antiparallel fashion. Sheets formed from such extended polypeptide chains may be stacked on top of one another.
5. Collagen fibrils are composed of extended polypeptide chains that are coiled in a left-handed manner. Three of these chains interact by hydrogen bonding and coil together into a right-handed cable. Collagen fibrils are composed of a staggered array of many such cables interacting in a side-by-side manner.
6. The structures of fibrous proteins are determined by the amino acid sequence, by the principle of forming the maximum number of hydrogen bonds, and by the steric limitations of the polypeptide chain, in which the peptide grouping is in a planar conformation.
7. X-ray diffraction provides data from which we can deduce the dimensions of the polypeptide chains in proteins. The use of x-ray techniques is, however, limited to molecules that can be oriented to achieve two- or three-dimensional order.
8. Fibrous proteins may achieve two-dimensional order, but they usually do not achieve three-dimensional order. Therefore, the diffraction pattern of fibrous proteins gives information about the regularly repeating elements along the long axis of the fibers but tells us very little about the orientation of amino acid side chains.
9. Many globular proteins can by crystallized to achieve three-dimensional order. Study of the crystals of a globular protein can lead to a complete determination of its three-dimensional structure.
10. The forces that hold globular proteins together are the same as those that hold fibrous proteins together, but there is less emphasis on regularity and more emphasis on burying the hydrophobic regions in the interior of the protein.
11. The secondary structures found in the keratins recur in smaller patches in globular proteins. Such regions of secondary structure are folded into a seemingly endless array of tertiary structures.

12. Tertiary structures can be understood in terms of a limited number of domains.
13. Quaternary structures are formed between nonidentical subunits to give irregular macromolecular complexes or between identical subunits to give geometrically regular structures.

Problems

1. The principal force driving the folding of some proteins is the movement of hydrophobic amino acid side chains out of an aqueous environment. Explain.

2. Outline the hierarchy of structural organization in proteins.

3. What is the role of loops or short segments of "random" structure in a protein whose structure is primarily α-helical?

4. What are some consequences of changing a hydrophilic residue to a hydrophobic residue on the surface of a globular protein? What are the consequences of changing an interior hydrophobic residue to a hydrophilic residue in the protein?

5. Some proteins are anchored to membranes by insertion of a segment of the N terminus into the hydrophobic interior of the membrane. Predict (guess) the probable structure of the sequence (Met-Ala-(Leu-Phe-Ala)$_3$-(Leu-Met-Phe)$_3$-Pro-Asn-Gly-Met-Leu-Phe). Why would this sequence by likely to insert into a membrane?

6. Suppose that every other Leu residue in the peptide shown in problem 5 was changed to Asp. Would this necessarily alter the secondary structure? Explain whether insertion into the membrane would be altered.

7. If you had several helical springs, how could you determine whether each spring was right- or left-handed?

8. "Left- and right-handed α helices of polyglycine are equally stable." Defend or refute this statement.

9. Urea

$$H_2N\overset{\overset{\displaystyle O}{\|}}{C}NH_2$$

and guanidinium chloride ($[(H_2N)_2C{=}NH_2]^+Cl^-$) are commonly used denaturants (cause loss of protein conformation). Provide explanations for how these substances might disrupt protein structures.

10. Many proteins (e.g., important metabolic enzymes) are insoluble in water and are found "attached" to membranes within cells. What amino acid residues do you expect to find on the "side" of the protein that "attaches" to the membrane?

11. Often the enzymes mentioned in problem 10 are purified with the aid of detergents such as sodium dodecylsulfate

$$CH_3(CH_2)_{11}{-}O{-}\overset{\displaystyle O}{\overset{\|}{\underset{\underset{\displaystyle O}{\|}}{S}}}{-}O^-\,{}^+Na$$

What is the function of the detergent?

12. It might be argued that in protein structure, as in everyday life, it is a "right-handed world." Use examples of protein structure discussed in this chapter to support this contention.

Solutions

1. In principle, proteins should assume a conformation yielding the lowest free-energy level. Entropic and enthalpic changes in the system (peptide plus surrounding medium) should sum to a negative value for the folding of the protein. Recall that $\Delta G = \Delta H - T\Delta S$. Thus, interactions that decrease ΔH or increase ΔS contribute to a more negative ΔG. Organized structure in the protein results in a decrease in entropy of the protein. This decrease in entropy must be offset by an increase in entropy in the surroundings. Removing hydrophobic residues from the aqueous interface is thermodynamically favorable and accounts, in part, for the increased entropy in solution surrounding the protein. Water molecules are more highly organized in the space immediately surrounding hydrophobic residues than in bulk water. Shifting the hydrophobic residues from an aqueous to an anhydrous environment decreases organization of the surrounding water and increases the entropic contribution to folding. In globular proteins dissolved in aqueous media, one finds that in general the hydrophobic residues are exposed on the surface of the protein to interact with water, whereas the hydrophobic residues coalesce in the core or interior of the protein. However, as noted in the text, enthalpically favorable interactions offset the decreased entropy of the folded protein and contributes to stabilization of the folded protein.
3. The α-helix is a rather rigid, rodlike secondary structural element that cannot easily change direction in space without breaking the helical arrangement. Frequently, *β* bends or loops, or in some instances segments of random coil structure, allow the helical elements to change direction and pack into a more compact globular structure. Myoglobin has a high helical content, but the segments of helix fold back on one another and pack into a globular protein.
5. The sequence given can assume an α-helical arrangement with the hydrophobic side chains located along the outside of the helix and exposed to solvent. The α-helix will be distorted at the proline residue and may enter a *β*-turn. The arrangement of the hydrophobic residues would

likely limit water solubility, because the increased organization of water structure surrounding the hydrophobic side chains decreases the entropic contribution to the stability of the system. However, the segment of the hydrophobic helix would be stabilized by insertion into the hydrophobic environment of the membrane. Moving the hydrophobic side chains on the helix from aqueous contact into the lipid bilayer would increase the entropic contribution of the water molecules to the system. Hydrophobic segments are found in proteins that are bound to biological membranes.

7. The right- or left-handedness of a helix is the same as a conventional screw or bolt. A right handed screw when turned clockwise, advances. The same is true of a helix or a helical spring.

9. When assessing how a specific substance might denature a protein it is often helpful to initially consider the forces/factors that stabilize and produce a protein's three dimensional shape: disulfide bonds, hydrophobic interactions, hydrogen bonds, and charge-charge (electrostatic) interactions. In addition the geometric constraints produced by the interactions of the multitude of individual component portions of a protein must be considered. Frequently denaturants function by providing an alternative to a portion of the protein in one or more of the above factors/forces. To determine how a denaturant might function, consider with which type of interaction mentioned above a denaturant might interact.

   Notice that both urea and guanidinium chloride can participate in several hydrogen bonds (both as the provider of the hydrogen and/or the lone pair of electrons). These latter substances can be visualized as literally inserting themselves into a hydrogen bond in the protein. This produces a urea or guanidinium between the components of the original hydrogen bond (forming two new hydrogen bonds) and causing a distortion in the shape of the protein. The reader should also notice that guanidinium is also a salt and can interact with electrostatic interactions in a manner comparable to that described for hydrogen bonds. Both urea and guanidinium chloride are rather small molecules.

11. Membrane bound enzymes are thought to typically have a cluster of hydrophobic residues on one portion of their surface. Presumably, these hydrophobic residues interact with the lipids present in the membrane. If membrane bound enzymes are separated from the membrane the total surface of the protein is exposed to the aqueous phase. Because of the number of surface hydrophobic residues these proteins are typically insoluble in water. If the hydrophobic portion of a detergent interacts with the hydrophobic surface region of the protein then the hydrophilic end of the detergent is free to interact with water, substantially increasing the water solubility of these proteins and allowing them to be purified free of the membrane.

5 Functional Diversity of Proteins: Hemoglobin and the Actin-Myosin Complex

Summary

In this chapter we considered the relationship between structural and functional properties for two protein systems.

1. Hemoglobin is a tetramer made of two almost identical subunits. The function of hemoglobin is threefold: to transport O_2 from the lungs to the tissues where it is consumed, to transport CO_2 from the tissues where it is produced to the lungs where it is expelled, and to maintain the blood pH over a narrow range. Cooperative interactions among the subunits allow hemoglobin to pick up the maximum amount of oxygen at high oxygen tensions in the lung tissue and to deliver the maximum amount of oxygen to the oxygen-consuming tissues.
2. Muscle is an aggregate of several different proteins. Its main protein components are organized as overlapping filaments of two types: thin filaments, composed mainly of actin molecules, and thick filaments, composed of myosin molecules. The process of muscular contraction entails a sliding of the two types of filaments past each other. In a fully contracted myofibril the actin and myosin filaments show a maximum overlap with each other. The contraction process involves the breakage and reformation of bridges between the actin and myosin molecules in a reaction that requires the expenditure of ATP.

Problems

1. In the later half of the 1800s one of the first suggestions that proteins were large molecules came from ashing experiments in which hemoglobin was converted to Fe_2O_3. These procedures suggested a molecular mass of hemoglobin greater than 15,900, a number unheard of at that time. Why did the researchers of the past century, who were excellent analytical chemists, deviate so significantly from the molecular mass of 64,500 now known for hemoglobin? What quantity of Fe_2O_3 results from the ashing of 1.00 g of hemoglobin?

2. Carbonic acid in the blood readily dissociates into hydrogen and bicarbonate ions. If the serum pH of 7.4 equals that inside the erythrocytes, what percentage of the carbonic acid is ionized? Use a value of 6.4 for the first pK_a of carbonic acid.

3. In addition to oxygen, hemoglobin subunits can also carry carbon dioxide. This is performed by covalent addition of CO_2 to the N terminal of the hemoglobin chains to produce a carbonate structure. Propose reactions for this process utilizing (a) CO_2 and (b) HCO_3^-.

4. Figure 5.13 indicates locations of mutations that have been shown to produce pathological conditions. The majority of the types of mutations that have been discovered in human hemoglobins have been mutations in which either amino acid residues bearing charges are replaced with ones with no charge or in which uncharged amino acids are replaced with charged amino acids. Do you think this represents a basic biological principle or is it an artifact of the detection process? Explain.

5. Sickle-cell anemia becomes most apparent during a sickle-cell crisis when the soft tissues are often acutely painful. Using information provided in the text on the molecular-cellular effects of the sickling of red blood cells, can you provide an explanation for the origin of the pain?

6. The hemoglobin present in a fetus is analogous to the $\alpha_2\beta_2$ tetramer of the adult, but the two β chains have been replaced with comparable γ chains. Considering the relevant biology, which hemoglobin type (adult or fetal) do you expect to have the greater affinity for oxygen?

7. Use the information presented in problem 6 above to propose a possible "future genetic engineering solution" to sickle-cell anemia. Are there any deleterious ramifications of your proposal?

8. Carbon 2 in glycerate-2,3-bisphosphate is in the D configuration. Would you expect the L configuration of glycerate-2,3-bisphosphate to have the same effect on the biochemistry of hemoglobin? Why or why not?

9. Figures 5.10 and 5.11 show a histidyl F8 residue interacting with the heme iron on deoxy- and oxyhemoglobin. Would you expect the imidazole nitrogen of the histidyl group to be protonated or unprotonated during this interaction? Why?

10. Would you expect the blood-hemoglobin system to transport more moles of O_2 or CO_2? Why?

11. The protein tropomyosin (TM) is composed of two identical chains of α helix (see table 5.1) that are in turn twisted around each other in a helical structure. Consider the average amino acid residue weight to be 105 daltons and use typical α-helix dimensions (fig. 4.4) to calculate the length of each chain in TM. Explain any discrepancy observed between the calculated length and the observed length of 360 Å.

12. Bryan Allen made aviation history by pedaling the *Gossamer Albatross* from near Folkestone, England, to Cap Gris Nez, France, from 4:51 AM to 7:40 AM on June 12, 1979. During this flight he continually produced about a third of a horsepower (*Nat. Geographic* 156:5,640, 1979). Considering that the energy available from the hydrolysis of ATP is 7.5 kcal/mole (see fig. 2.9), determine the number of moles of ATP required for this flight. For simplicity, assume that the muscles are 50% efficient in the conversion of chemical energy into mechanical energy and that one horsepower equals the energy expenditure of 178 cal/s.

13. What is the cause of rigor after death?

14. Explain how muscular contraction is regulated.

Solutions

1. The early chemists had no way of knowing each hemoglobin (Hb) contains four Fe^{2+}.

$$\frac{1\text{ g Hb}}{} \left| \frac{1\text{ mole Hb}}{64{,}500\text{ g Hb}} \right| \frac{4\text{ mole Fe}}{1\text{ mole Hb}} \left| \frac{1\text{ mole }Fe_2O_3}{2\text{ mole Fe}} \right| \frac{160\text{ g }Fe_2O_3}{1\text{ mole }Fe_2O_3}$$

$$\frac{1000\text{ mg}}{\text{g}} = 4.9\text{ mg }Fe_2O_3$$

3. a. CO_2 + H_2N-hemoglobin → $^{-}OC(=O)HN$-hemoglobin + H^+

b. $HOCO_2^-$ + H_2N-hemoglobin → $^{-}OC(=O)HN$-hemoglobin + H_2O

5. During a sickle cell crisis the red blood cells collapse from their normal "disk" shape into a sickle (or quarter moon) shape. This collapsed shape has a smaller cross sectional area than the original "disk" shape. In the capillaries the red blood cells normally pass in a single-file. If the cells are sickled it is possible for them to wedge together, clogging the capillary, stopping the flow of red blood cells, and depriving the adjacent tissues of oxygen. The oxygen deprivation produces the pain during a crisis.
7. Individuals with sickle cell anemia have a functional gene for the gamma chain. If the production of fetal hemoglobin could be "turned back on" the affected individuals could function normally except in pregnancy. During pregnancy, net transfer of oxygen from the mother to the fetus would be inhibited due to the fact that both mother and fetus would have the same fetal hemoglobin with equal affinity for oxygen.
9. The interaction between the histidyl F8 residue and heme iron is a ligand-metal ion interaction. In this complexation reaction the ligand provides a lone pair of electrons. Both the protonated and unprotonated imidizol ring of histidine have a lone pair(s) of electrons. The protonated form of histidyl F8 is positively charged which is less likely to interact favorably with the Fe^{2+} of the heme. Also note that during coordinate covalent bond formation the imidizol nitrogen takes on a formal positive charge, which would be unfavorable in a protonated (already positive) histidyl group. If the unprotonated form is utilized then the lone pair which complexes the iron is in an sp^2 orbital. The coordinate covalent bond is then in conjugation with the imidazol ring, allowing resonance.

11. $$\frac{32{,}000}{} \left| \frac{\text{AA}}{105\text{ D}} \right| \frac{\text{turn}}{3.6\text{AA}} \left| \frac{5.4\text{ Å}}{\text{turn}} \right. = 457\text{ Å}$$

Because the two 457 Å long strands are twisted into a helix the resulting TM is shorter than 457 Å.

13. After death, muscles enter a stage of rigor in which they are fully contracted and the maximum number of bridges are formed. Rigor is probably due to the depletion of ATP and a considerable discharge of calcium from the sarcoplasmic reticulum. In living tissue, the cytosolic Ca^{2+} concentration is restored to resting levels within 30 ms of receiving a signal, and the myofibrils relax.

6 Methods for Characterization and Purification of Proteins

Summary

In this chapter we considered the different means of purification that enable us to study individual proteins in isolation. The following points are the most important.

1. Whereas proteins must be studied *in vivo* in their normal habitat, to characterize them in great detail they must also be isolated in pure form. Protein purification is a complex art, and a great variety of purification methods are usually applied in sequence for the purification of any protein.
2. Frequently the first step in protein purification is differential sedimentation of broken cell parts. In this way soluble proteins may be separated from organelle-sequestered proteins.
3. Following differential sedimentation, proteins may be separated into crude fractions by the addition of increasing amounts of $(NH_4)_2SO_4$. Specific proteins characteristically precipitate in a limited range of salt concentrations.
4. Column procedures are useful in fractionating proteins with different affinity properties and sizes. By the use of different column materials in conjunction with specific eluting solutions, highly purified protein preparations can be obtained.
5. Gel electrophoresis, which separates proteins according to their size and charge, can be used in purification or more frequently as a means of assaying the purity during a purification. Isoelectric focusing, which separates proteins according to their isoelectric points, can be used for the same purpose.
6. The molecular weights of soluble proteins can be roughly estimated by SDS gel electrophoresis or can be rigorously determined by sedimentation techniques.
7. The purification of two proteins, UMP synthase from mammalian tumor cells, and lactose carrier protein from *E. coli* bacteria is described in detail to illustrate how different fractionation methods can be combined most effectively.

Problems

1. Why are “salting out” procedures often used as an initial purification step following the production of a crude extract by centrifugation?

2. Rarely is DEAE-cellulose used above a pH of about 8.5. Can you provide a reason(s) why?

3. Given that the only structural difference between phosphorylase a and phosphorylase b is that phosphorylase a has a covalently bound phosphate on serine 14, do you expect phosphorylase a and phosphorylase b to elute as a single peak on DEAE-cellulose chromatography? What if gel filtration was utilized?

4. A method for the purification of 6-phosphogluconate dehydrogenase from *E. coli* is summarized in the table. For each step, calculate the specific activity, percentage yield, and degree of purification (*n*-fold). Indicate which step results in the greatest purification. Assume that the protein is pure after gel (Bio-Gel A) exclusion chromatography. What percentage of the initial crude cell extract protein was 6-phosphogluconate dehydrogenase?

Step	**Volume (ml)**	**Total Protein (mg)**	**Total Units**	**Specific Activity (U/mg)**	**Yield (%)**	**Purification (*n*-fold)**
Cell extract	2,800	70,000	2,700			
$((NH_4)_2SO_4)$ fractionation	3,000	25,400	2,300			
Heat treatment	3,000	16,500	1,980			
DEAE chromatography	80	390	1,680			
CM-cellulose	50	47	1,350			
Bio-Gel A	7	35	1,120			

5. Although used effectively in the 6-phosphogluconate dehydrogenase isolation procedure, heat treatment cannot be used in the isolation of all enzymes. Explain.

6. Assume that the isoelectric point (pI) of 6-phosphogluconate dehydrogenase is 6. Explain why the buffer used in the DEAE-cellulose chromatography must have a pH greater than 6 but less than 9 for the enzyme to bind to the DEAE resin.

7. Will the 6-phosphogluconate dehydrogenase bind to the CM-cellulose in the same buffer pH range used with the DEAE-cellulose? Explain. In what pH range would you expect the dehydrogenase to bind to CM-cellulose? Explain.

8. Examine the isolation procedure shown in problem 4 and explain why gel exclusion chromatography is used as the final step rather than as the step following the heat treatment.

9. A student isolated an enzyme from anaerobic bacteria and subjected a sample of the protein to SDS polyacrylamide gel electrophoresis. A single band was observed on staining the gel for protein. His adviser was excited about the result, but suggested that the protein be subjected to electrophoresis under nondenaturing (native) conditions. Electrophoresis under nondenaturing conditions revealed two bands after the gel was stained for protein. Assuming the sample had not been mishandled, offer an explanation for the observations.

10. A salt-precipitated fraction of ribonuclease contained two contaminating protein bands in addition to the ribonuclease. Further studies showed that the one contaminant had a molecular weight of about 13,000 (similar to ribonuclease) but an isoelectric point 4 pH units more acidic than the pI of ribonuclease. The second contaminant had an isoelectric point similar to ribonuclease but had a molecular weight of 75,000. Suggest an efficient protocol for the separation of the ribonuclease from the contaminating proteins.

11. You have a mixture of proteins with the following properties:

Protein 1: M_r 12,000, pI = 10
Protein 2: M_r 62,000, pI = 4
Protein 3: M_r 28,000, pI = 8
Protein 4: M_r 9,000, pI = 5

Predict the order of emergence of these proteins when a mixture of the four is chromatographed in the following systems:

a. DEAE-cellulose at pH 7, with a linear salt gradient elution.
b. CM-cellulose at pH 7, with a linear salt gradient elution.
c. A gel exclusion column with a fractionation range of 1,000–30,000 M_r, at pH 7.

12. You wish to purify an ATP-binding enzyme from a crude extract that contains several contaminating proteins. To purify the enzyme rapidly and to the highest purity, you must consider some sophisticated strategies, among them affinity chromatography. Explain how affinity chromatography can be applied to this separation, and explain the physical basis of the separation.

Solutions

1. The "salting out" procedure can be done on essentially any volume of material. It is often a rapid effective way to reduce the volume of crude extracts and at the same time eliminate a major portion of the total protein. The speed/simplicity of the process is also important in that it often eliminates proteolytic enzymes which can hydrolyze the desired proteins.
3. There is essentially no difference in the size (mass) of phosphorylase a and phosphorylase b; therefore, they should elute as a single peak upon gel filtration. Phosphorylase a has a bound phosphate which would give it a greater negative charge than phosphorylase b at pH's greater than about 2 (pK_{a1} of a phosphate ester). This difference in charge allows the separation of phosphorylase a from phosphorylase b with the use of DEAE-cellulose.
5. Heat treatment of protein solutions denatures and precipitates some of the proteins, while others remain both soluble and stable. Thermal lability is determined empirically for each enzyme or protein of interest. The enzyme or protein of interest may be rapidly denatured by the thermal treatment, necessitating alternative strategies for isolation.
7. The ion-exchanger group on the CM-cellulose is a weak carboxylic acid whose pK is around 4. Above pH 4, the carboxylic acid is unprotonated and negatively charged. However, in the buffer system given in problem 3, the protein is negatively charged and will not adhere to the negatively charged CM-cellulose. The protein is positively charged at pH more acidic than the pI. Therefore the protein will probably adhere to (bind to) the CM-cellulose if the pH is between 4 and 6.
9. Proteins are separated by SDS-PAGE on the basis of molecular weight of the denatured protein. The molecular weight is the minimum value or subunit molecular weight. Two proteins that differ in some properties but have the same subunit molecular weight will likely appear as a single protein band after SDS-PAGE. Thus, a 40,000 M_r dimer composed of 2 × 20,000 M_r subunits will not be separated from an 80,000 M_r tetramer composed of 4 × 20,000 M_r subunits.

Native, or nondenaturing, electrophoresis separates proteins based on mass/charge characteristics. The student's "pure" protein contains at least two components separable by the criterion of mass/charge but which share a common subunit molecular weight. It is also possible that the multiple protein bands appearing in the nondenaturing gel arose through deamidation of glutamine or asparagine residue side chains.

11. a. DEAE cellulose will have a net positive charge at pH 7 due to protonation of the weakly basic tertiary amine (pK_a approximately 8.5). The exchanger will bind or retard negatively charged proteins. The charge on each protein may be estimated from the pI. At pH values greater than pI, the protein will be negatively charged. At pH below pI, the protein will be positively charged.

Protein	pI	Charge at pH 7
Protein 1	10	(+)
Protein 2	4	(–)
Protein 3	8	(+)
Protein 4	5	(–)

Proteins 1 and 3 should elute in the initial wash buffer, but proteins 2 and 4 are predicted to bind to the column. Based solely on isoelectric point, one might predict that protein 4, then protein 2, would be eluted in the salt gradient.

b. The carboxyl group of the CM-cellulose will be negatively charged at pH 7 because of deprotonation of the weak acid (pK_a approximately 4). The charge on each protein was established in the previous section.

Proteins 2 and 4 would be eluted in the initial wash buffer from the column, whereas proteins 1 and 3 would be predicted to adhere to the column. Based solely on pI values, protein 3 would be predicted to elute prior to protein 1 in the KCl gradient. In each separation of proteins based on ion exchange, the actual elution order must be determined experimentally. Proteins having the same or similar pI but differing in the absolute number of charged residues will likely differ in their elution characteristics.

c. Gel exclusion chromatography separates globular proteins on the basis of relative size, which is a function of molecular weight. Larger proteins (higher molecular weight) elute before smaller (lower molecular weight) proteins. The gel exclusion resin proposed to separate the four proteins has an exclusion limit of 30,000. Proteins with molecular weight greater than the limit (protein 2: 62,000 M_r) are excluded from entry into the gel and elute in the void volume (V_o). The other proteins in the solution will elute in the order protein 3, protein 1, protein 4.

7 Enzyme Kinetics

Summary

Enzymes are biological catalysts. Kinetic analysis is one of the most broadly used tools for characterizing enzymatic reactions.

1. The rate of a reaction depends on the frequency of collisions between the reacting species and on the fraction of the collisions that produce products. The former depends on the concentrations of the reactants; the latter depends on temperature and activation free energy $\Delta G^{\ddagger}$. $\Delta G^{\ddagger}$ can be interpreted as the free energy needed to convert the reactants to a transition state. A catalyst increases the reaction rate by lowering $\Delta G^{\ddagger}$.
2. Enzyme kinetics usually are studied by mixing the enzyme and substrates and measuring the initial rate of formation of product or the disappearance of a reactant. Special techniques are necessary to measure very fast reactions. It is common to measure the rate as a function of substrate concentration, pH, and temperature.
3. Enzymes have localized catalytic sites. The substrate (S) binds at the active site to form an enzyme-substrate complex (ES). Subsequent steps transform the bound substrate into product and regenerate the free enzyme. The overall speed of the reaction depends on the concentration of ES. Shortly after the enzyme and substrate are mixed, [ES] becomes approximately constant and remains so for a period of time termed the steady state. The rate (v) of the reaction in the steady state usually has a hyperbolic dependence on the substrate concentration. It is proportional to [S] at low concentrations but approaches a maximum (V_{max}) when the enzyme is fully charged with substrate. The Michaelis constant K_m is the substrate concentration at which the rate is half maximal. K_m and V_{max} often can be obtained from a plot of $1/v$ versus $1/[S]$. If ES is in equilibrium with the free enzyme and substrate, K_m is equal to the dissociation constant for the complex (K_s). More generally, K_m depends on at least three rate constants and is larger than K_s.
4. The turnover number k_{cat} is the maximum number of molecules of substrate converted to product per unit time per active site and is V_{max} divided by the total enzyme concentration. The specificity constant k_{cat}/K_m is a measure of how rapidly an enzyme can work at low substrate concentrations. This is usually the best index of the effectiveness of an enzyme.
5. Enzymes that catalyze reactions of two or more substrates work in a variety of ways that can be distinguished by kinetic analysis. Some enzymes bind their substrates in a fixed order; others bind in random order. In some cases binding of one substrate gives a partial reaction before the second substrate binds.
6. Enzymes can be inhibited by agents that interfere with the binding of substrate or with conversion of the ES complex into products. Reversible inhibitors are classified as competitive, noncompetitive or uncompetitive. A competitive inhibitor competes with substrate for binding at the active site. Consequently, a sufficiently high concentration of substrate can eliminate the effect of a competitive inhibitor. Noncompetitive inhibitors bind at a separate site and block the reaction regardless of whether the active site is occupied by substrate. An uncompetitive inhibitor binds to the ES complex but not to the free enzyme. These three forms of inhibition

are distinguishable by measuring the rate as a function of the concentrations of the substrate and inhibitor. Irreversible inhibitors often provide information on the active site by forming covalently linked complexes that can be characterized.

Problems

1. Explain what is meant by the order of a reaction, using the reaction below as an example. What is the reaction order for each reactant? For the overall reaction? (Consider the forward and reverse reaction.)

$$A + B \rightleftharpoons 2C$$

2. In a first-order reaction a substrate is converted to product so that 87% of the substrate is converted in 7 min. Calculate the first-order rate constant. In what time is 50% of the substrate converted to product?

3. Prove that the K_m equals the substrate concentration at one-half maximal velocity.

4. The Michaelis constant K_m is frequently equated with K_s, the [ES] dissociation constant. However, there is usually a disparity between those values. Why? Under what conditions are K_m and K_s equivalent?

5. When quantifying the activity of an enzyme, does it matter if you measure the appearance of a product or the disappearance of a reactant?

6. An enzyme was assayed with substrate concentration of twice the K_m value. The progress curve of the enzyme (product produced per minute) is shown here. Give two possible reasons why the progress curve becomes nonlinear.

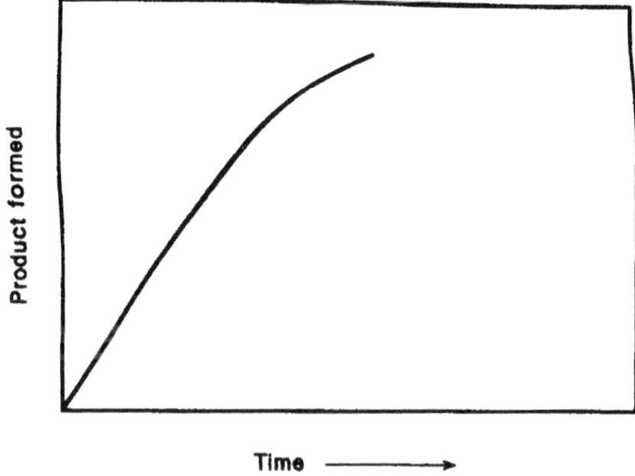

7. What is the steady-state approximation and under what conditions is it valid?

8. Assume that an enzyme-catalyzed reaction follows Michaelis-Menten kinetics with a K_m of 1 μM. The initial velocity is 0.1 μM/min at 10 mM substrate. Calculate the initial velocity at 1 mM, 10 μM, and 1 μM substrate. If the substrate concentration increased to 20 mM, would the initial velocity double? Why or why not?

9. If the K_m for an enzyme is 1.0×10^{-5} M and the K_i of a competitive inhibitor of the enzyme is 1.0×10^{-6} M, what concentration of inhibitor would be necessary to lower the reaction rate by a factor of 10 when the substrate concentration is 1.0×10^{-3} M? 1.0×10^{-5} M? 1.0×10^{-6} M?

10. Assume that an enzyme-catalyzed reaction follows the scheme shown:

$$E + S \underset{k_2}{\overset{k_1}{\rightleftharpoons}} ES \underset{k_4}{\overset{k_3}{\rightleftharpoons}} E + P$$

Where $k_1 = 10^9\ \text{M}^{-1}\ \text{s}^{-1}$, $k_2 = 10^5\ \text{s}^{-1}$, $k_3 = 10^2\ \text{s}^{-1}$, $k_4 = 10^7\ \text{M}^{-1}\ \text{s}^{-1}$, and $[E_t]$ is 0.1 nM. Determine the value of each of the following.

a. K_m
b. V_{max}
c. Turnover number
d. Initial velocity when $[S]_o$ is 20 μM.

11. A colleague has measured the enzymatic activity as a function of reaction temperature and obtained the data shown in this graph. He insists on labeling point A as the "temperature optimum" for the enzyme. Try, tactfully, to point out the fallacy of that interpretation.

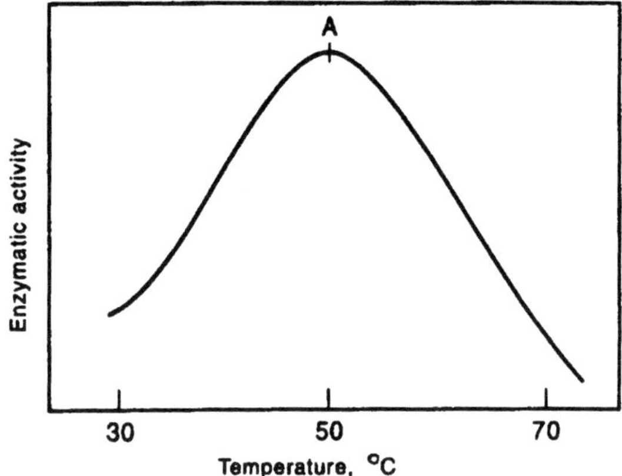

12. You have isolated a tetrameric NAD^+-dependent dehydrogenase. You incubate this enzyme with iodoacetamide in the absence or presence of NADH (at 10 times the K_m concentration), and you periodically remove aliquots of the enzyme for activity measurements and amino acid composition analysis. The results of the analyses are shown in the table.

	(No NADH Present)				(NADH Present)			
Time (min)	*Activity (U/mg)*	*His*	*(Residues/mole)*	*Cys*	*Activity (U/mg)*	*His*	*(Residues/mole)*	*Cys*
0	1,000	20		12	1,000	20		12
15	560	18.2		11.4	975	20		11.4
30	320	17.3		10.8	950	20		10.8
45	180	16.7		10.4	925	19.8		10.4
60	100	16.4		10.0	900	19.6		10.0

a. What can you conclude about the reactivities of the cysteinyl and histidyl residues of the protein?
b. Which residue can you implicate in the active site? On what do you base the choice? Are the data conclusive concerning the assignment of a residue to the active site? Why or why not?
c. After 1 h you dilute the enzyme incubated with iodoacetamide but no NADH. Do you expect the enzyme activity to be restored? Explain.

13. The initial velocity data shown in the table were obtained for an enzyme.

[S] (mM)	Velocity (Ms^{-1}) × 10^7
0.10	0.96
0.125	1.12
0.167	1.35
0.250	1.66
0.50	2.22
1.0	2.63

Each assay at the indicated substrate concentration was initiated by adding enzyme to a final concentraton of 0.01 nM. Derive K_m, V_{max}, k_{cat}, and the specificity constant.

14. You measured the initial velocity of an enzyme in the absence of inhibitor and with inhibitor A or inhibitor B. In each case, the inhibitor is present at 10 μM. The data are shown in the table.

[S] (mM)	Velocity (Ms^{-1}) × 10^7 Uninhibited	Velocity (Ms^{-1}) × 10^7 Inhibitor A	Velocity (Ms^{-1}) × 10^7 Inhibitor B
0.333	1.65	1.05	0.794
0.40	1.86	1.21	0.893
0.50	2.13	1.43	1.02
0.666	2.49	1.74	1.19
1.0	2.99	2.22	1.43
2.0	3.72	3.08	1.79

a. Determine K_m and V_{max} of the enzyme.
b. Determine the type of inhibition imposed by inhibitor A and calculate K_i(s).
c. Determine the type of inhibition imposed by inhibitor B and calculate K_i(s).

Solutions

1. Reaction order is the power to which a reactant concentration is raised in defining the rate equation. The rate equation describing the forward reaction in the example is

$$v_f = k(A)(B)$$

where (A) and (B) are expressed as molar concentration and k has the units $M^{-1}s^{-1}$. The units of v_f are $M\ s^{-1}$. The concentration of substrates A and B are each raised to the first power in the rate equation. The reaction is first order in A and in B. The overall reaction order is the sum of the order contributions of each reactant and is second order.

The rate equation for the reverse reaction is

$$v_r = k_2(C)^2$$

3. Let $v = V_{max}/2$ and substitute into Equation 25.

$$\frac{V_{max}}{2} = \frac{V_{max}[S]}{[S]+K_M}, \quad \text{solving gives: } K_M = [S]$$

5. Normally the chemical substance that is the easiest to quantify is chosen and it does not matter if it is the appearance of a product or disappearance of a reactant that is determined.
7. Steady-state approximation is based on the concept that the formation of [ES] complex by binding of substrate to free enzyme, and breakdown of [ES] to form product plus free enzyme occur at equal rates. A graphical representation of the relative concentrations of free enzyme, substrate, enzyme-substrate complex, and product is shown in figure 7.8 in the text. Note that [S] is initially much larger than $[E_t]$. If we assume

$$E + S \underset{k_2}{\overset{k_1}{\rightleftharpoons}} ES \xrightarrow{k_3} E + P$$

and that [P] is zero, the rate of formation of [ES] is expressed as

$$v_f = k_1[E][S]$$

and the rate of breakdown is expressed as

$$v_r = k_2[ES] + k_3[ES]$$
$$v_r = (k_2 + k_3)\,[ES]$$

At steady state, the two velocities v_f and v_r are equal. One can apply the distribution (or conservation) expression ($E_t = [E] + [ES]$) and the kinetic constants to arrive at the Michaelis-Menten expression, an expression derived using the steady-state assumption.

$$v = \frac{V_{max}[S]}{K_m + [S]}$$

where $V_{max} = k_3\,[E_t]$ and $K_m = (k_2 + k_3)/(k_1)$.
Steady-state approximation may be assumed until the substrate concentration is depleted with a concomitant decrease in the concentration of [ES].

9. Enzyme velocity in the presence of a competitive inhibitor is expressed as

$$v_i = \frac{V_{max}[S]}{K_m\left(1 + [I]/K_i\right) + [S]}$$

where K_i is the inhibitor dissociation constant. Let v be the velocity of the reaction in the absence of the inhibitor. If v is 100%, the inhibitor concentration yielding v_i = 10% v is sought. Now,

$$\frac{v}{v_i} = \frac{V_{max}[S]}{K_m + [S]} \times \frac{K_m\left(1 + [I]/K_i\right) + [S]}{V_{max}[S]}$$

$v/v_i = 10$ and the expression simplifies:

$$10(K_m + [S]) = K_m + K_m/K_i\,[I] + [S]$$
$$9(K_m + [S]) = K_m/K_i\,[I]$$

Substitute the known values and derive the expression

$$9(10^{-5}\text{M} + [S]) = 10^{-5}/10^{-6}[I]$$
$$[I] = 0.9(10^{-5}\text{M} + [S])$$

The amount of inhibitor required to inhibit the reaction 90% may now be calculated.

When $[S] = 1.0 \times 10^{-3}\text{M},\ [I] = 9.1 \times 10^{-4}\text{M}$
$[S] = 1.0 \times 10^{-5}\text{M},\ [I] = 1.8 \times 10^{-5}\text{M}$
$[S] = 1.0 \times 10^{-6}\text{M},\ [I] = 9.9 \times 10^{-6}\text{M}$

Thus, increasing the concentration of substrate at a fixed inhibitor concentration yields less inhibition.

11. The rate of a chemical reaction increases with temperature as defined by the Arrhenius expression (see chapter 7, pg. 138 in the text). Since an enzyme is a catalyst for a chemical reaction, the rate of an enzyme catalyzed reaction increases with increased temperature. However, the catalyst, a protein is structurally labile and is inactivated (denatured) at elevated temperatures. The precise temperature at which the enzyme is inactivated varies with the specific enzyme. The figure in problem 11 illustrates the expected increase in reaction rate with increased temperature until the temperature at point A is reached. The temperature at point A roughly approximates the maximum temperature at which the catalyst (enzyme) is stable. Denaturation or inactivation removes the catalyst from the reaction and the reaction rate decreases because the observed velocity is dependent on the concentration of enzyme. There is no "temperature optimum" for a catalyst (enzyme).

13. The kinetic data are analyzed by graphing the reciprocal of the velocity as a function of the reciprocal of the substrate concentration. The kinetic constants V_{max} and K_m are obtained from the resulting Lineweaver Burk plot.

Substrate (mM^{-1})	Velocity ($\text{sec M}^{-1} \times 10^{-7}$)
10	1.04
8	0.89
6	0.74
4	0.60
2	0.45
1	0.38

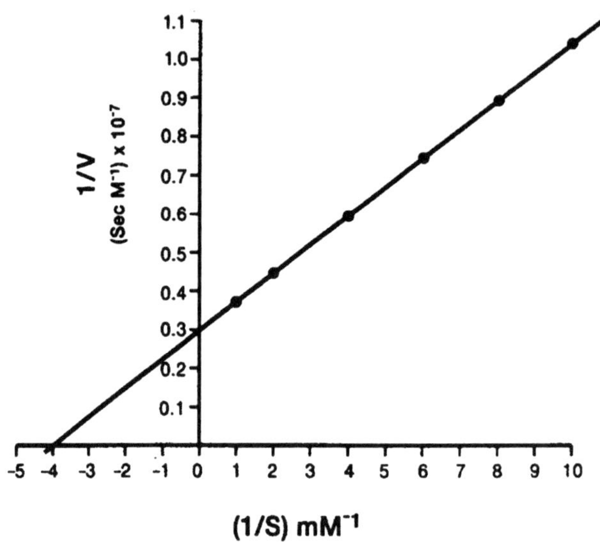

The intercept on the negative X axis is the value of $(-1/K_m)$.

$$-(K_m)^{-1} = -4.17 \text{ mM}^{-1}$$

$$K_m = 0.24 \text{ mM} = (2.4 \times 10^{-4} \text{ M})$$

The Y intercept is $(V_{max})^{-1}$.

$$(V_{max})^{-1} = 0.305$$

$$V_{max} = 3.3 \times 10^{-7} \text{ M s}^{-1}$$

Turnover number (k_{cat} or k_3) = $V_{max}/[E_t]$

$$k_{cat} = (3.3 \times 10^{-7} \text{ M s}^{-1})/(10^{-11}\text{M}) = 3.3 \times 10^{4} \text{ s}^{-1}$$

Specificity constant = k_{cat}/K_m

$$= (3.3 \times 10^{4})/(2.4 \times 10^{-4})$$

$$= 1.4 \times 10^{8} \text{ M}^{-1} \text{ s}^{-1}$$

8 How Enzymes Work

Summary

Like ordinary chemical catalysts, enzymes interact with the reacting species in a manner that lowers the free energy of the transition state.

1. All known enzymatic reaction mechanisms depend on one or more of the following five themes:
   a. proximity effects (enzymes hold the reactants close together in an appropriate orientation)
   b. general-acid or general-base catalysis (acidic or basic groups of the enzyme donate or remove protons and often do first one and then the other)
   c. electrostatic effects (charged or polar groups of the enzyme favor the redistribution of electric charges that must occur to convert the substrate into the transition state)
   d. nucleophilic or electrophilic catalysis (nucleophilic or electrophilic functional groups of the enzyme or a cofactor react with complementary groups of the substrate to form covalently linked intermediates
   e. structural flexibility (changes in the protein structure can increase the specificity of enzymatic reactions by ensuring that substrates bind or react in an obligatory order and by sequestering bound substrates in pockets that are protected from the solvent)
2. The serine proteases trypsin, chymotrypsin, and elastase are very similar in structure but have substrate-binding pockets that are tailored for different amino acid side chains. The active site of each enzyme contains three critical residues: serine, histidine, and aspartate, which are positioned so that the serine hydroxyl group becomes a strong nucleophilic reagent for reaction with the substrate's peptide carbon atom. This reaction generates an acyl-enzyme intermediate. Formation and hydrolysis of the acyl-enzyme intermediate probably are promoted by electrostatic interactions that stabilize tetrahedral transition states and by general-acid and general-base catalysis.
3. Ribonuclease A hydrolyzes RNA adjacent to pyrimidine bases. The reaction proceeds through a 2′,3′-phosphate cyclic diester intermediate. Formation and breakdown of the cyclic diester appear to be promoted by concerted general-base and general-acid catalysis by two histidine residues, and by electrostatic interactions with two lysines. These reactions proceed through pentavalent phosphoryl intermediates. The geometry of these intermediates resembles the geometry of vanadate compounds that act as inhibitors of the enzyme.
4. Triosephosphate isomerase interconverts dihydroxyacetone phosphate and 3-phosphoglyceraldehyde. A glutamic acid residue probably acts as a general base to remove a proton from the substrate, forming an enediolate intermediate. Histidine and lysine side chains stabilize this intermediate electrostatically. Triosephosphate isomerase appears to have reached evolutionary perfection in the sense that it catalyzes its reaction at the maximum possible rate given the concentration of the substrate in the cell.

Problems

1. Hexokinase promotes the following reaction:

$$\text{Glucose} + \text{ATP} \rightarrow \text{ADP} + \text{Glucose-6-phosphate}$$

Hexokinase initially binds glucose, then the hexokinase-glucose complex binds ATP. How do you explain the purpose of this substrate-binding pattern to a fellow student without using the induced-fit concept?

2. a. In what ways are the mechanistic features of chymotrypsin, trypsin, and elastase similar?
   b. If the mechanisms of these enzymes are similar, what features of the enzyme active site dictate substrate specificity?

3. If a lysine were substituted for the aspartate in the trypsin side-chain-binding crevice, would you expect the enzyme to be functional? If it were functional, what effect would you predict the substitution to have on substrate specificity?

4. Plant proteolytic enzymes are cysteine proteases; that is, they have a cysteine that is critical for the catalytic activity. Plant proteases are thought to function by a mechanism reminiscent of that shown for chymotrypsin (fig. 8.11). Propose a structure for the acyl-enzyme intermediate that would exist for plant proteases.

5. How could the inhibitors presented in table 7.4 be used to indicate that an unknown protease was of plant versus nonplant origin?

6. Notice how the active site of chymotrypsin contains Asp 102, His 57, and Ser 195 (fig. 8.11), whereas the active site of RNase A contains Asp 121, Lys 41, Lys 7, His 12, and His 119 (fig. 8.18). What fundamental principle concerning protein structure is emphasized when the residue number in the active site of enzymes is considered?

7. RNase A utilizes a 2′,3′-cyclic phosphate ester as an intermediate (fig. 8.18) but yields only a 3′-phosphate product. Within the mechanism, what controls the final position of the phosphate (i.e., why isn't the 2′-phosphate a product)?

8. For many enzymes, V_{max} is dependent on pH. At what pH do you expect V_{max} of RNase to be optimal? Why?

9. RNase can be completely denatured by boiling or by treatment with chaotropic agents (e.g., urea), yet can refold to its fully active form on cooling or removal of the denaturant. By contrast, when enzymes of the trypsin family and carboxypeptidase A are denatured, they do not regain full activity on renaturation. What aspects of trypsin and carboxypeptidase A structure preclude their renaturation to the fully active form?

10. In figures 8.18 and 8.21 lysyl residues are shown in the lower right parts of the figures. An initial examination of the mechanisms indicates these lysyl groups are only observers. Why are they shown in the "mechanisms" and what do they do?

11. Why do structural analogs of the transition-state intermediate of an enzyme inhibit the enzyme competitively and with low K_i values?

12. Transition-state analogs of a specific chemical reaction have been used to elicit antibodies with catalytic activity. These catalytic antibodies have great promise as experimental tools as well as having commercial value. Why is it reasonable to assume that the binding site for the transition-state analog on the antibody mimics the enzyme active site? What difficulties might be encountered if a catalytic antibody is sought for a reaction requiring a cofactor (coenzyme)?

13. Using site-directed mutagenesis techniques, you isolate a series of recombinant enzymes in which specific lysine residues are replaced with aspartate residues. The enzymatic assay results are shown in the table.

Enzyme Form	Activity (U/mg)
Native enzyme	1,000
Recombinant Lys 21→ Asp 21	970
Recombinant Lys 86 → Asp 86	100
Recombinant Lys 101 → Asp 101	970

a. What do you infer about the role(s) of Lys 21, 86, and 101 in the catalytic mechanism of the native enzyme?
b. Speculate on the location of Lys 21 and Lys 101. Would you expect these residues to be conserved in an evolutionary sense?
c. Would you expect Lys 86 to be evolutionarily conserved? Why or why not?

Solutions

1. The substrate recognition site is a pocket on hexokinase into which glucose, then ATP, bind. Glucose and hexokinase bind together, changing the shape of both forming a binding site for ATP.
3. Trypsin cleaves peptide bonds on the carboxyl side of arginine or lysine residues. This specificity is achieved through specific interaction of the positively charged side chains with an aspartic acid residue located at the base of the substrate side chain binding pocket. The aspartate in the binding crevice is not to be confused with Asp 102, a component of the catalytic

triad. Substitution of lysine for the aspartate in the substrate binding pocket should have no effect on the ability of the enzyme to hydrolyze peptide bonds. The catalytic triad (Asp-His-Ser) would remain functional assuming that substitution caused minimal disruption of the protein conformation. Specificity of bond cleavage would most certainly change. The favorable electrostatic interaction between the Lys or Arg side chain of the peptide substrate and the aspartate in the binding pocket would be replaced by electrostatic repulsion upon substitution of the lysine residue. Thus, selective binding of Arg or Lys to the substrate pocket would be precluded. The modified trypsin might therefore (a) cleave peptide bonds randomly, but exclude bonds on the carboxyl side of Lys or Arg residues, or (b) exhibit specificity for peptide cleavage on the carboxyl side of acidic amino acids (Asp, Glu). Side chains of these amino acids may fit well into the substrate binding pocket and have favorable electrostatic interaction with the lysine therein.

5. Because aminal proteases rely on a serine to form an alkoxide nucleophile and non-animal proteases use a cysteine to form a thioalkoxide nucleophile, table 7.4 can be consulted to fine reagents that react with seryl or cysteinyl side chains. Iodoacetate and para-mercuribenzoate react with thiols so they would be likely to inhibit plant porteases while dilsopropyl-fluorophosphate which reacts with seryl side chains (specifically those with a more labile alcohol hydrogen) would be likely to inhibit animal proteases.

7. Histidyl 12 accepts the 2′-OH hydrogen to start the reaction. The protonated histidyl 12 ultimately provides the hydrogen to the cyclic ester to produce the 3′-phosphate product. Presumably the relative position of the histidyl 12 and ribose ring have not changed during the reaction.

9. Proteins are assembled from amino acids into the primary sequence based on information encoded in the gene sequence. The primary sequence dictates the formation of secondary structures, and these structures fold and interact to form the tertiary structure. Renaturation of denatured protein is also dictated by the primary structure of the protein.

   The trypsin family of enzymes and carboxypeptidase A are each synthesized as proenzymes that are proteolytically activated after synthesis. The proteolyzed, active enzymes have primary structures that differ from the gene product (proenzyme) and would not be expected to refold into a conformation equivalent to that of the proenzyme. It is not surprising that the renatured enzyme is not catalytically active. In addition, zinc is a cofactor required for carboxypeptidase A activity and must be available during the renaturation process. Addition of Zn^{2+} to the renaturing proenzyme would likely result in the correct ligand attachment to the metal. The correct placement of the ligands may not occur in the refolding of denatured, proteolytically activated enzymes.

11. The enzyme active site stabilizes the transition-state intermediate formed during catalysis with a resulting decrease in the activation energy of the reaction. The structure of the transition-state intermediate is complementary to the structure of the active site and is thus bound much more tightly than are either substrates or products. Transition-state analogs, having structural features only slightly different from the transition-state intermediate, would be expected to bind very tightly to the active site and thus have a low K_i. The analog would compete with substrate or product for the active site and would exhibit competitive inhibition.

13a. Recombinant enzyme in which lysine residues at positions 21 and 101 were replaced with aspartic acid residues (Lys 21 → Asp 21 and Lys 101 → Asp 101) exhibited only 3% less maximal catalytic activity than did the nonrecombinant (native) enzyme. Therefore, these residues were neither functional active-site residues, nor were they critical for maintaining a catalytically competent conformation of the enzyme. Recombinant enzyme in which Lys 86 was replaced with Asp 86 exhibited catalytic activity that was only 10% that of the nonrecombinant enzyme, suggesting that Lys 86 is a critical residue either at or near the catalytic active center or is essential for maintaining a catalytically competent conformation of the enzyme.

13b. Lysins 21 and 101 are likely outside the catalytic site and may not be evolutionarily conserved. However, that assessment is based solely on measurement of catalytic activity *in vitro*. Amino acid residues on the surface of proteins, a location reasonably assumed for these residues, have other roles that may be physiologically relevant: (a) as part of structural motifs recognized by regulatory enzymes, (b) as one of a cluster of charged amino acid side chains important in assembly of multienzyme complexes, or (c) in interaction of the enzyme with the surface of the membrane.

13c. Lysine 86 is required for maximum enzymatic activity and would likely be conserved. Amino acids whose side chains are involved either with the binding of substrate or cofactors, chelating transition metals, as reactants in the catalytic chemistry, or as essential structural determinants are usually strongly conserved among groups of enzymes.

9 Regulation of Enzyme Activities

Summary

Cells regulate their metabolic activities by controlling rates of enzyme synthesis and degradation and by adjusting the activities of specific enzymes. Enzyme activities vary in response to changes in pH, temperature, and the concentrations of substrates or products, but also can be controlled by covalent modifications of the protein or by interactions with activators or inhibitors.

1. Partial proteolysis, an irreversible process, is used to activate proteases and other digestive enzymes after their secretion and to switch on enzymes that cause blood coagulation. Common types of reversible covalent modification include phosphorylation, adenylylation, and disulfide reduction.
2. Allosteric effectors are inhibitors or activators that bind to enzymes at sites distinct from the active sites. Allosteric regulation allows cells to adjust enzyme activities rapidly and reversibly in response to changes in the concentrations of substances that are structurally unrelated to the substrates or products. The initial steps in a biosynthetic pathway commonly are inhibited by the end products of the pathway, and numerous enzymes are regulated by ATP, ADP, or AMP.
3. The kinetics of allosteric enzymes typically show a sigmoidal dependence on substrate concentration rather than the more common hyperbolic saturation curves. These enzymes usually have multiple subunits, and the sigmoidal kinetics can be ascribed to cooperative interactions of the subunits. Binding of substrate to one subunit changes the dissociation constant for substrate on another subunit. The extent of the cooperativity can be described by the Hill equation or by the "symmetry" model or the more general "sequential" model. The symmetry model postulates that the enzyme can exist in two conformations, T and R. It is assumed that the substrate binds more tightly to the R conformation than to the T conformation, that binding of the substrate or an allosteric effector changes the equilibrium between these conformations, and that cooperative interactions make all of the subunits switch from one conformation to the other in a concerted manner. The symmetry model provides a useful conceptual framework that is consistent with the behavior of many allosteric enzymes.
4. Phosphofructokinase, the key regulatory enzyme of glycolysis, has four identical subunits. It exhibits sigmoidal kinetics with respect to fructose-6-phosphate and is inhibited by ATP and stimulated by ADP. In the transition between the R and T conformations, the subunits rotate with respect to each other and there is a rearrangement of the binding site for fructose-6-phosphate, which is located at an interface between subunits. At another interface, antiparallel β strands of two subunits are hydrogen-bonded together in the T structure but are separated by a row of water molecules in the R structure. This structural feature explains the cooperative nature of the conformational transition in all of the subunits. Intermediate structures probably would be less stable than either the R or the T structure.
5. Aspartate carbamoyl transferase, the first enzyme in the biosynthesis of pyrimidines, is inhibited allosterically by CTP, an end product of the pathway, and is stimulated by ATP. It has six identical catalytic (c) subunits and six regulatory (r) subunits. The c and r subunits can be separated by treating the enzyme with mercurials. Binding of a substrate analog to one of the c subunits causes the entire c_6r_6 complex to flip to the R conformation; binding of CTP to an r

subunit favors the T conformation. Again, the substrate-binding sites are located at interfaces between subunits, and the T to R transition results in a rotation of the subunits and brings together components of the active site.

6. Glycogen phosphorylase breaks down glycogen to glucose-1-phosphate. It exists in two forms that differ by a covalent modification. Phosphorylase *a* is phosphorylated on a serine residue and is the more active form under cellular conditions. Phosphorylase *b*, which lacks the phosphate, is stimulated allosterically by AMP but inhibited by ATP. The enzyme has two identical subunits, with the binding site for allosteric effectors at the interface. In the T to R transition there is a repositioning of arginine and aspartate residues at the binding site for P_i. Conversion of phosphorylase *b* to phosphorylase *a* tightens the interactions between the subunits and favors the transition to the R form.

Problems

1. Notice the salt bond in figure 9.1b between the beta carboxyl group of Asp 194 and the N-terminus (Ile 16) in the structure of chymotrypsin. Why is the N-terminus amino acid number 16?

2. Many hemophiliacs are sustained by regular injections of one of the cascade components, for example, Factor VIII (fig. 9.2). Why doesn't the addition of Factor VIII cause an uncontrolled clotting of the blood in these individuals?

3. While reading this chapter you should have noticed that protein regulation mechanisms fall into two major categories: Covalent modifications and allosteric regulation. Does one approach seem to be inherently less complicated? Why?

4. What similarities and differences can you find in the covalent modification of proteins shown in figures 9.3 and 9.4?

5. The cAMP-dependent protein kinases phosphyorylate specific Ser (Thr) residues on target proteins. Given the availability of serine and threonine residues on the surface of globular proteins, how might a protein kinase select the "correct" residues to phosphorylate?

6. Is the interaction between an allosteric effector and an allosteric enzyme always an equilibrium?

7. The substrate concentration yielding half-maximal velocity is equal to K_m for hyperbolically responding enzymes. Is this relationship true for allosterically or sigmoidally responding enzymes? How do positive or negative allosteric effectors change the substrate concentration required for half-maximal velocity?

8. ATP is both a substrate and an inhibitor of the enzyme phosphofructokinase (PFK). Although the substrate fructose-6-phosphate binds cooperatively to the active site, ATP does not bind cooperatively. Explain how ATP may be both a substrate and an inhibitor of PFK.

9. Aspartate carbamoyltransferase is an allosteric enzyme in which the active sites and the allosteric effector binding sites are on different subunits. Explain how it might be possible for an allosteric enzyme to have both kinds of sites on the same subunit.

10. Examine the relationship of aspartate carbamoyltransferase (ACTase) activity to aspartate concentration shown in the figure. Estimate the ($S_{0.9}/S_{0.1}$) ratio for the reaction under the following conditions: (a) normal curve, (b) plus 0.2 mM CTP, (c) plus 0.8 mM ATP. Do these ratios differ significantly? Explain.

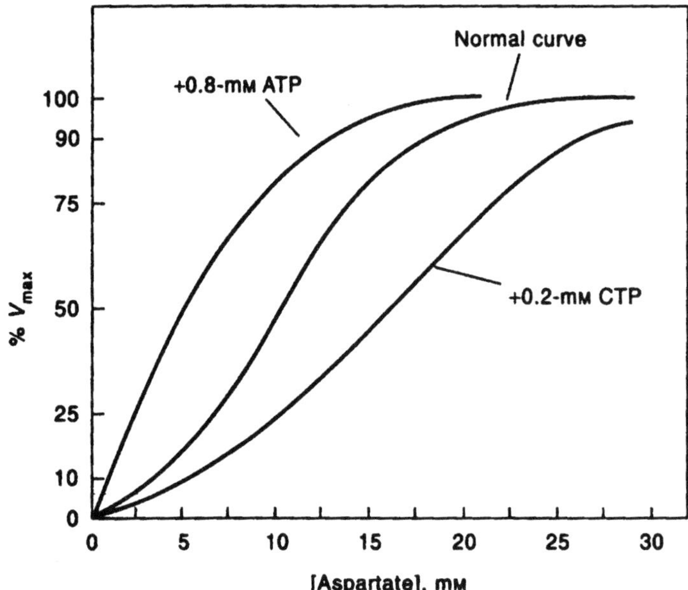

11. Why do you think the researchers who initially prepared PALA (fig. 9.16) utilized a methylene between the carbonyl and phosphoryl group? Why didn't they "just" use a "regular" phosphate group. *Hint:* Compare the postulated reaction intermediate (fig. 9.16) with the reaction shown in figure 9.13.

12. In some instances, protein kinases are activated or inhibited by low-molecular-weight modifiers. Explain how a metabolite might more effectively regulate an enzyme by modifying a protein kinase rather than directly inhibiting the target enzyme.

Solutions

1. The Ile in chymotrypsinogen is the 16th amino acid. During the conversion to chymotrypsin a 15 amino acid peptide is cleaved from chymotrypsinogen. The original amino acid number system was retained in chymotrypsin, making the N-terminus amino acid number 16.
3. Many students visualize the allosteric control model as a simpler control mechanism. The activity of allostene proteins (enzyme) is controlled by the interaction (equilibrium) of an enzyme binding site with a small molecule. The response time and number components involved is minimal. In contrast, covalent modification requires an enzyme to perform the modification, a cofactor such as ATP, and often an enzyme to reverse the original modification. The ezyme(s) involved in the modification and any reversal of the modification must be controlled at some point (e.g., at the transcription level).
5. Phosphorylation of serine or threonine (and possibly tyrosine) on the target protein may be largely influenced by the amino acid sequence around these residues. These amino acid sequences may define a specific motif or recognition site for the protein kinase. Kemp *(TIBS.* 15:342–346 (1990)) and Kennelly and Krebs (*J. Biol. Chem.* 266:15555–58 (1991)) have reviewed possible recognition sequences (motifs) on the substrates for a number of protein kinases. The authors caution that not all serine (threonine) residues in a specific motif are necessarily phosphorylated. Serine (or threonine) residues may not be phosphorylated because of other factors, such as topographical features on the target protein, that prevent binding of the target protein to the protein kinase active site.
7. The substrate concentration required for half-maximal activity ($S_{0.5}$) of an allosterically regulated enzyme will depend on the relationship between enzyme reaction velocity and substrate concentration as affected by activators and inhibitors, if present. Hence ($S_{0.5}$) may be decreased with allosteric activators and may be increased with allosteric inhibitors. At a given substrate concentration, the activity of an allosterically regulated enzyme may vary widely and depend on the presence of other modifiers (activators or inhibitors). This feature makes these enzymes well suited for their role in metabolic regulation.
9. Allosteric regulation of an enzyme having a binding site for a regulatory molecule and an active site on the same subunit is not uncommon. Regulatory molecules bind at sites separate from the active site and induce a conformational change that affects substrate binding to the active site on the same as well as adjacent subunits.
11. The methylene group prevents the elimination of a phosphate, which is required to convert the postulated intermediate into carbamoyl-aspartate. The suggested oxygen analog might eliminate phosphate (i.e., be a substrate for aspartate carbamoyl transferase).

10 Vitamins and Coenzymes

Summary

Coenzymes are molecules that act in cooperation with enzymes to catalyze biochemical processes, performing functions that enzymes are otherwise chemically not equipped to carry out. Most coenzymes are derivatives of the water-soluble vitamins, but a few, such as hemes, lipoic acid, and iron-sulfur clusters, are biosynthesized in the body. Each coenzyme plays a unique chemical role in the enzymatic processes of living cells.

1. Thiamine pyrophosphate promotes the decarboxylation of α-keto acids and the cleavage of α-hydroxy ketones.
2. Pyridoxal-5′-phosphate promotes decarboxylations, racemizations, transaminations, aldol cleavages, and elimination reactions of amino acid substrates.
3. Nicotinamide coenzymes act as intracellular electron carriers to transport reducing equivalents between metabolic intermediates. They are cosubstrates in most of the biological redox reactions of alcohols and carbonyl compounds and also act as cocatalysts with some enzymes.
4. Flavin coenzymes act as cocatalysts with enzymes in a large number of redox reactions, many of which involve O_2.
5. Phosphopantetheine coenzymes form thioester linkages with acyl groups, which they activate for group transfer reactions.
6. Lipoic acid mediates electron transfer and acyl-group transfer in α-keto acid dehydrogenase complexes.
7. Biotin mediates carboxylation of activated methyl groups.
8. Phosphopantetheine, lipoic acid, and biotin, by virtue of their long, flexible structures, facilitate the physical translocation of chemically reactive species among separate catalytic sites.
9. Tetrahydrofolates are cosubstrates for a variety of one-carbon transfer reactions. Tetrahydrofolates maintain formaldehyde and formate in chemically poised states, making them available for essential processes by specific enzymatic action.
10. Heme coenzymes participate in a variety of electron-transfer reactions, including reactions of peroxides and O_2. Iron-sulfur clusters, composed of Fe and S in equal numbers with cysteinyl side chains of proteins, mediate other electron-transfer processes, including the reduction of N_2 to 2 NH_3. Nicotinamide, flavin, and heme coenzymes act cooperatively with iron-sulfur proteins in multienzyme systems that catalyze hydroxylations of hydrocarbons and also in the transport of electrons from foodstuffs to O_2.
11. Ascorbic acid is needed as a reductant to maintain some enzymes in their active forms.
12. Metal ions serve as catalytic elements in some enzymes; Zn^{2+} is particularly important in this regard. Ca^{2+} binds tightly to some proteins and acts to trigger intracellular responses to hormonal signals.
13. In general, less is known about the mechanisms of action of the lipid-soluble vitamins than about the coenzymes derived from water-soluble vitamins. The structures and functions of vitamins D, K, E, and A are discussed briefly.

Problems

1. Which of the coenzymes listed in table 10.1 can be considered to be derivatives of AMP?

2. What structural features of biotin and lipoic acid allow these cofactors to be covalently bound to a specific protein in a multienzyme complex yet participate in reactions at active sites on other enzymes of the complex?

3. The following reactions are catalyzed by pyridoxal-5′-phosphate-dependent enzymes. Write a reaction mechanism for each, showing how pyridoxal-5′-phosphate is involved in catalysis.

a. $$CH_3{-}\overset{\overset{\displaystyle O}{\|}}{C}{-}COO^- + R{-}\underset{\underset{\displaystyle NH_3^+}{|}}{CH}{-}COO^- \longrightarrow CH_3{-}\underset{\underset{\displaystyle NH_3^+}{|}}{CH}{-}COO^- + R{-}\overset{\overset{\displaystyle O}{\|}}{C}{-}COO^-$$

b. $$H_3\overset{+}{N}{-}(CH_2)_4{-}\underset{\underset{\displaystyle NH_3^+}{|}}{CH}{-}CO_2^- \longrightarrow CO_2 + H_3\overset{+}{N}{-}(CH_2)_5{-}NH_3^+$$

c. $$^-O_2C{-}CH_2{-}\underset{\underset{\displaystyle NH_3^+}{|}}{CH}{-}CO_2^- \longrightarrow CO_2 + CH_3{-}\underset{\underset{\displaystyle NH_3^+}{|}}{CH}{-}CO_2^-$$

4. $NADP^+$ differs from NAD^+ only by phosphorylation of the C-2′ OH group on the adenosyl moiety. The redox potentials differ only by about 5 mV. Why do you suppose it is necessary for the cell to employ two such similar redox cofactors?

5. Thiamine-pyrophosphate-dependent enzymes catalyze the reactions shown below. Write a chemical mechanism that shows the catalytic role of the coenzyme.

a.

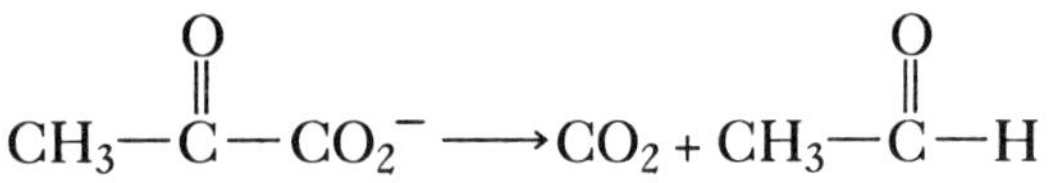

$$CH_3-\overset{\overset{O}{\|}}{C}-CO_2^- \longrightarrow CO_2 + CH_3-\overset{\overset{O}{\|}}{C}-H$$

b.

$$\text{Ⓟ}OCH_2-(CHOH)_2-\overset{\overset{OH}{|}}{CH}-\overset{\overset{O}{\|}}{C}-CH_2OH + HOPO_3^{2-} \longrightarrow$$

$$\text{Ⓟ}O-CH_2-(CHOH)_2-CHO + CH_3-\overset{\overset{O}{\|}}{C}-OPO_3^{2-} + H_2O$$

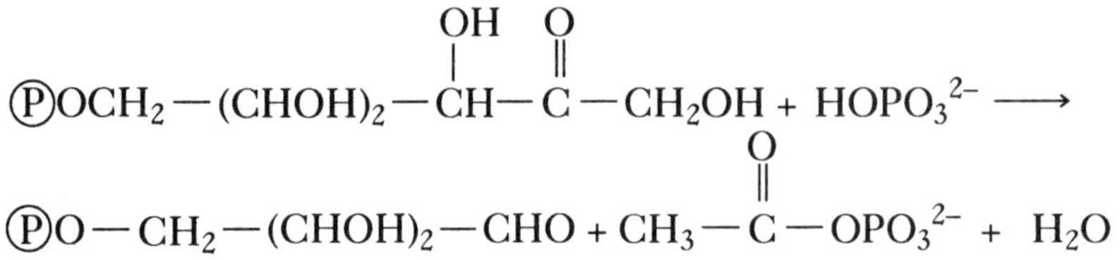

6. What chemical features allow flavins (FAD, FMN) to mediate electron transfer from NAD(P)H to cytochromes or iron-sulfur proteins?

7. What coenzymes would you expect to participate in the following reactions (specifically indicate which are enzyme-bound coenzymes or "substrate/product-like" coenzymes).

a.

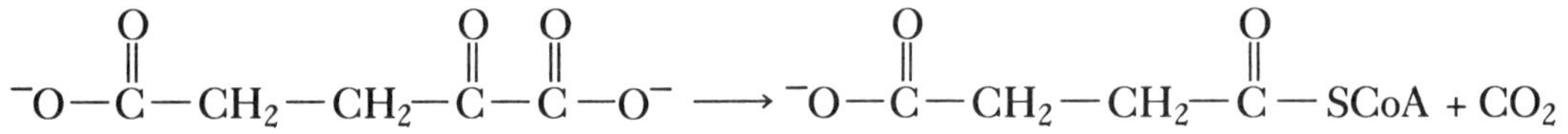

$$^-O-\overset{\overset{O}{\|}}{C}-CH_2-CH_2-\overset{\overset{O}{\|}}{C}-\overset{\overset{O}{\|}}{C}-O^- \longrightarrow {}^-O-\overset{\overset{O}{\|}}{C}-CH_2-CH_2-\overset{\overset{O}{\|}}{C}-SCoA + CO_2$$

b.

$$CH_3-CH_2-\overset{\overset{O}{\|}}{C}-SCoA + ATP + HCO_3^- \longrightarrow CH_3\underset{\underset{\underset{\underset{O^-}{|}}{O=C}}{|}}{\overset{\overset{H}{|}}{C}}-\overset{\overset{O}{\|}}{C}-SCoA + ADP + HOPO_3^{2-}$$

c.

$$CH_3-CH_2-CH_2-\overset{\overset{O}{\|}}{C}-SCoA \longrightarrow CH_3-CH=CH-\overset{\overset{O}{\|}}{C}-SCoA$$

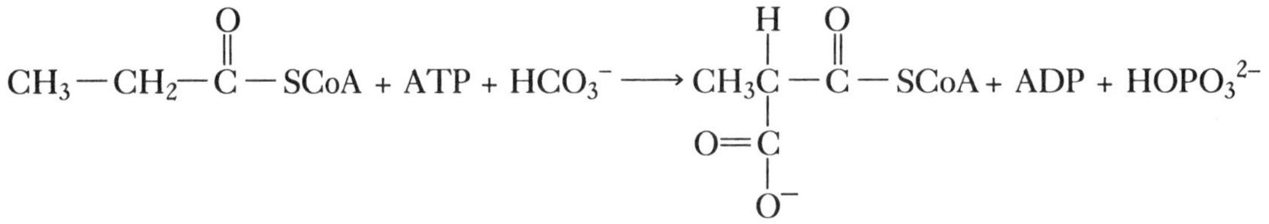

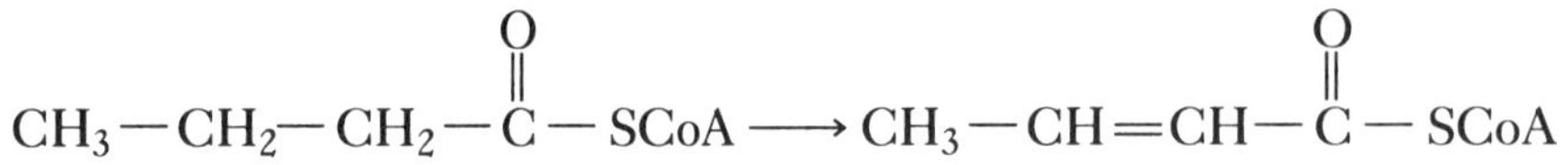

8. Write mechanisms that indicate the involvement of biotin in the following reactions:

a.

$$CH_3-\overset{\overset{\large O}{\|}}{C}-SCoA + HCO_3^- + ATP \longrightarrow {}^-O_2C-CH_2-\overset{\overset{\large O}{\|}}{C}-SCoA + ADP + HOPO_3^{2-}$$

b.

$$CH_3-CH_2-\overset{\overset{\large O}{\|}}{C}-SCoA + {}^-O_2C-CH_2-\overset{\overset{\large O}{\|}}{C}-CO_2^- \rightleftharpoons$$

$${}^-O_2C-\underset{\underset{\large CH_3}{|}}{CH}-\overset{\overset{\large O}{\|}}{C}-SCoA + CH_3-\overset{\overset{\large O}{\|}}{C}-CO_2^-$$

9. a. What metabolic advantage is gained by having flavin cofactors covalently or tightly bound to the enzyme?
   b. Would covalently bound NAD^+ ($NADP^+$) be a metabolic advantage or disadvantage?

10. Given the amino acid of the general structure

$$CH_3-\underset{\underset{\large X}{|}}{CH}-\underset{\underset{\large NH_3^+}{|}}{CH}-COO^-$$

we could use pyridoxal-5′-phosphate to eliminate X, decarboxylate the amino acid, or oxidize the α carbon to a carbonyl with formation of pyridoxamine-5′-phosphate. The metabolic diversity afforded by PLP, unchanneled, could wreak havoc in the cell. What other components are required to channel the PLP-dependent reaction along specific reaction pathways?

11. What coenzymes covered in this chapter are (a) biological redox agents, (b) acyl carriers, (c) both redox agents and acyl carriers?

12. For each of the following enzymatically catalyzed reactions, identify the coenzyme involved:

a.

$$R-\underset{\substack{|\\ NH_3^+}}{CH}-COO^- + 2\,H^+ + O_2 \longrightarrow R-\overset{\substack{O\\ \|}}{C}-COO^- + NH_4^+ + H_2O_2$$

b.

$$HO-CH_2-\underset{\substack{|\\ NH_3^+}}{CH}-CO_2^- \longrightarrow CH_3-\overset{\substack{O\\ \|}}{C}-CO_2^- + NH_4^+$$

c.

$$CH_3-CH_2-\overset{\substack{O\\ \|}}{C}-SCoA + HCO_3^- + ATP \longrightarrow {}^-O_2C-\underset{\substack{|\\ CH_3}}{CH}-\overset{\substack{O\\ \|}}{C}-SCoA + ADP + P_i$$

d.

$$\underset{\substack{|\\ O(P)}}{CH_2}-\underset{\substack{|\\ OH}}{CH}-\underset{\substack{|\\ OH}}{CH}-\overset{\substack{OH\\ |}}{CH}-\overset{\substack{O\\ \|}}{C}-\overset{\substack{OH\\ |}}{CH_2} + \underset{\substack{|\\ O(P)}}{CH_2}-\underset{\substack{|\\ OH}}{CH}-CHO \rightleftharpoons$$

$$\underset{\substack{|\\ O(P)}}{CH_2}-\underset{\substack{|\\ OH}}{CH}-\underset{\substack{|\\ OH}}{CH}-CHO + \underset{\substack{|\\ O(P)}}{CH_2}-\underset{\substack{|\\ OH}}{CH}-\overset{\substack{OH\\ |}}{CH}-\overset{\substack{O\\ \|}}{C}-\underset{\substack{|\\ OH}}{CH_2}$$

13. The structure of ascorbic acid is shown in figure 10.16. Which hydrogen is the most acidic? Why?

14. Some bacterial toxins use NAD^+ as a true substrate rather than as a coenzyme. The toxins catalyze the transfer of ADP-ribose to an acceptor protein. Examine the structure of NAD^+ and indicate which portion of the molecule is transferred to the protein. What is the other product of the reaction?

Solutions

1. The following coenzymes contain the AMP moiety: NAD^+, NADH, $NADP^+$, NADPH, FAD, $FADH_2$, and CoASH.

3. a.

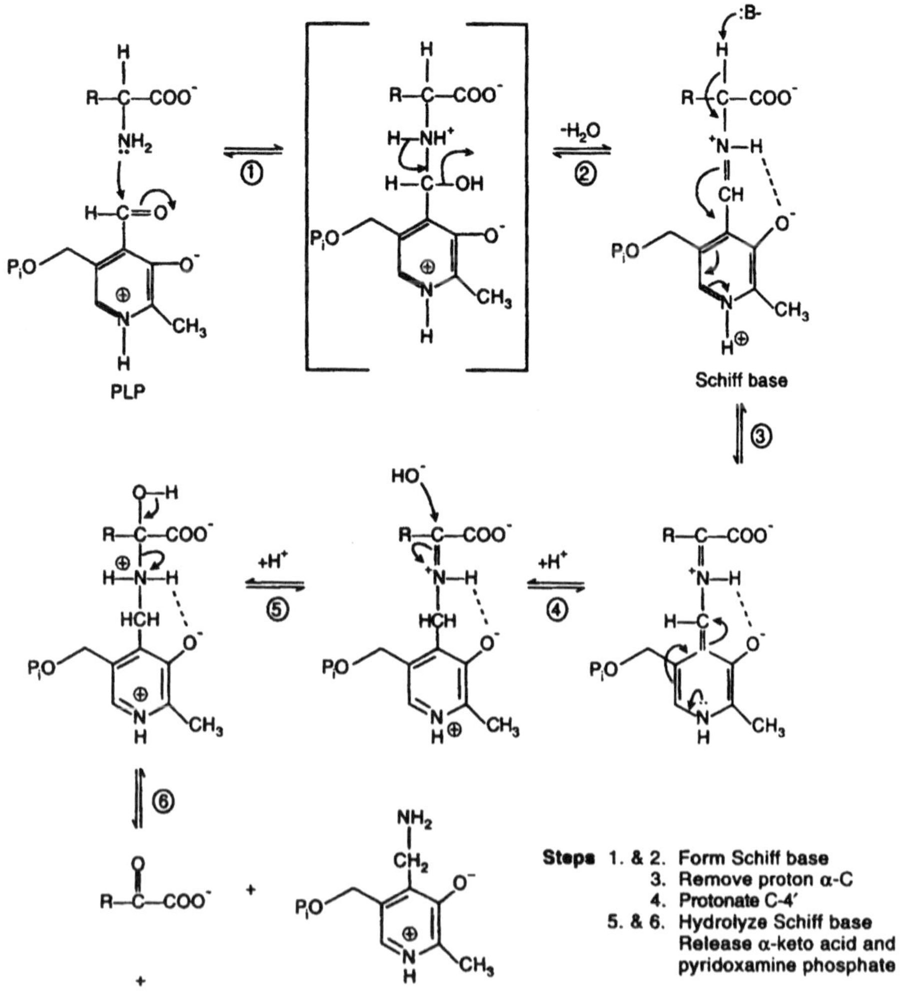

PLP
Schiff base
-H₂O
+H⁺
Steps 1. & 2. Form Schiff base
3. Remove proton α-C
4. Protonate C-4′
5. & 6. Hydrolyze Schiff base
Release α-keto acid and pyridoxamine phosphate

Transfer of amino group from pyridoxamine phosphate to pyruvate, forming alanine occurs by reversal of the steps shown above. Other transaminases use other α-amino acids and α-keto acids.

b.

$H_3^+N-(CH_2)_4-C(H)(-N^+H=CH-\text{pyridoxal phosphate})-COO^-$ $\xrightarrow[①]{-CO_2}$ $H_3^+N-(CH_2)_4-CH=N^+H-\ldots$ $\xrightarrow{②}$ $H_3^+N-(CH_2)_4-CH_2-N^+H=CH-\ldots$

$\xrightarrow[③]{HO^-}$ $[H_3^+N-(CH_2)_4-CH_2-N^+H_2-CH(OH)-\ldots]$ $\xrightarrow{④}$ $P_iO-\text{pyridoxal (CHO)}$ + $^+NH_3-(CH_2)_5-NH_3^+$

Steps 1. Decarboxylation of α-COO^-
2. Protonate α-Carbon
3. & 4. Hydrolyze Schiff base

c.

① $\xrightarrow{-\alpha H^+}$ ② $\xrightarrow[\text{C-4'}]{\text{Protonate}}$ ③ $\xrightarrow{-\beta\ COO^-}$ ④ $\rightleftharpoons$ (Protonate β-Carbon) ⑤ $\xrightarrow[\text{C-4'}]{-H^+}$ ⑥ $\xrightarrow[\alpha C]{+H^+}$ ⑦ $\xrightarrow[\text{Schiff base}]{\text{Hydrolyze}}$ CH_3—CH($NH_3^{\oplus}$)—COO^- + PLP

5. a.

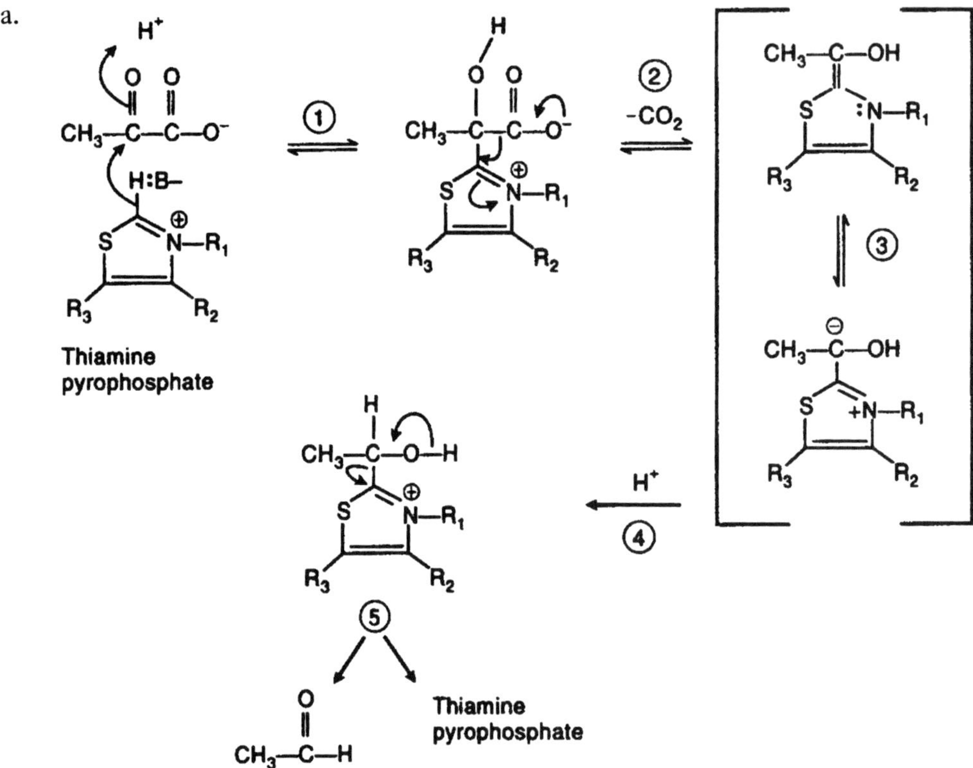

Steps
1. Deprotonation of TPP to ylid form, nucleophilic addition of ylid to α-ketogroup
2. Decarboxylation
3. Resonance stabilization
4. Protonation, elimination of TPP

b.

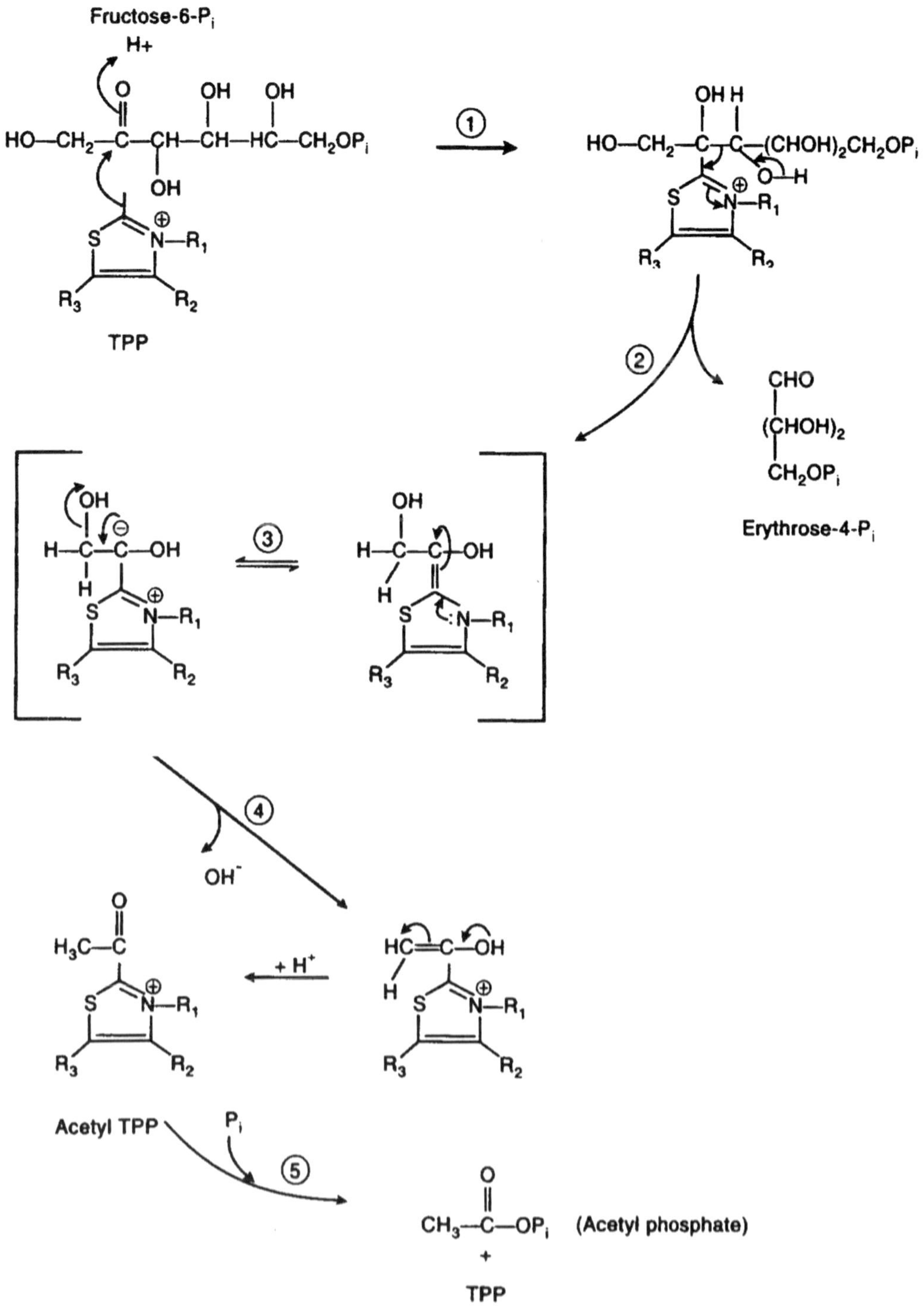

Fructose-6-P_i

H+

$HO-CH_2-C(=O)-CH(OH)-CH(OH)-HC(OH)-CH_2OP_i$

TPP

① → $HO-CH_2-C(OH)(TPP)-C(H)(OH)-(CHOH)_2CH_2OP_i$

②

CHO
$(CHOH)_2$
CH_2OP_i

Erythrose-4-P_i

③

④

OH^-

$+ H^+$

Acetyl TPP

P_i

⑤

$CH_3-C(=O)-OP_i$ (Acetyl phosphate)

+

TPP

Steps

1. Deprotonation to ylid form, nucleophilic addition of ylid to fructose-6-phosphate
2. Oxidation of C-3, release of erythrose-4-phosphate
3. Resonance stabilization of intermediate
4. Elimination of -OH from C-2
5. Transfer of acyl group to phosphate

7. a. thiamine pyrophosphate, lipoic acid, FAD, NAD^+, CoASH (an alpha ketoacid dehydrogenase).
   b. biotin.
   c. FAD.
9. a. Reduced flavins in solution are rapidly reoxidized by molecular oxygen forming superoxide radical and H_2O_2, metabolites of oxygen that are toxic to the cell. Were free FAD or FMN used in biological transfer, for example in the mitochondrial electron system, oxygen could compete with other oxidants for the reduced flavin and disrupt energy transduction. If the flavin component of mitochondrial succinate dehydrogenase were directly reoxidized by molecular oxygen, rather than by ubiquinunt (see chapter 15 in the text), oxidative phosphorylation driven by succinate oxidation would be abolished. However, sequential transfer of the reducing equivalents via the electron transport chain to cytochrome oxidase drives phosphorylation of approximately 1.5 moles ADP per mole of succinate with the reduction of O_2 to H_2O rather than H_2O_2. Enzyme-bound flavins are usually shielded from uncontrolled oxidation by O_2.
   b. Tightly bound $NAD(P)^+$ is an advantage to enzymes catalyzing rapid H:—removal and readdition to a substrate in a stereospecific fashion. Freely diffusing NAD(P)H is an advantage in transferring reducing equivalents among catalytic sites of various dehydrogenases. NAD(P)H in solution is not rapidly oxidized by molecular oxygen as is soluble FADH or FMNH and is suited in its role as a freely diffusing redox reagent in the cell.
11. a. redox agents: FAD, FMN, NAD^+, $NADP^+$, and lipoyl.
    b. acyl carriers: CoASH, lipoyl, and thiamine pyrophosphate.
    c. both acyl carriers and redox agents: lipoyl.
13. The hydroxyl hydrogen on carbon 3 is the most acidic because the resulting anion is resonance stabilized to a greater degree than the hydroxyl hydrogen on carbon 2.

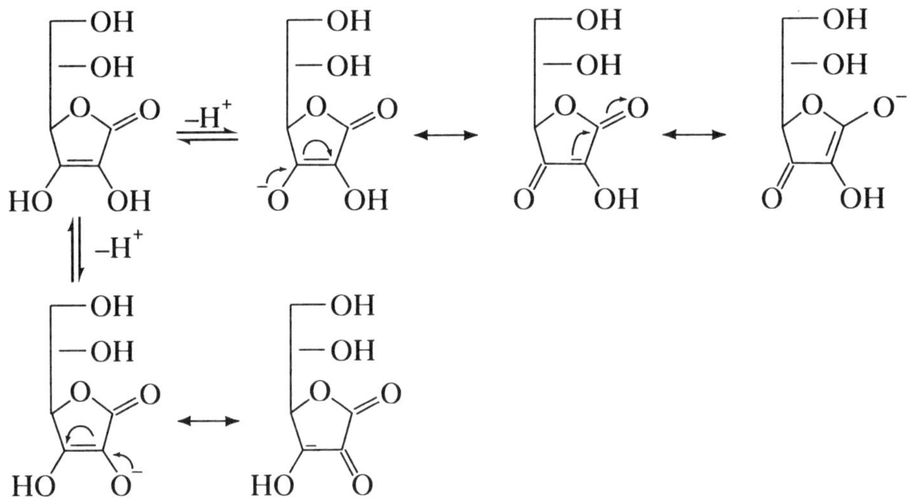

11 Metabolic Strategies

Summary

1. All cells need energy and starting materials for synthesis. Ultimately these are supplied by autotrophic organisms, especially green plants; in plants the starting materials are made from CO_2 and the supply of chemical energy and reducing power is dependent on the absorption of light energy. In a heterotrophic organism, the role of catabolism is to supply those basic needs from conversions (usually oxidation) of foodstuffs.
2. Metabolic chemistry is characterized by functionality. Each reaction is important because of its participation in a sequence of reactions, and each sequence interacts with other sequences.
3. Sequences may be broadly classified into two main types: biosynthetic (anabolic) and degradative (catabolic). Anabolic sequences usually require energy, and catabolic sequences usually produce energy.
4. Metabolic regulatory mechanisms have evolved so as to stabilize concentrations of key metabolites under a broad range of conditions.
5. Energy is coupled from catabolic sequences of energy-requiring activities of a cell by the ATP-ADP system. In a similar manner, reducing power is coupled by the NADPH-$NADP^+$ system.
6. The stoichiometry of coupling to ATP-to-ADP conversions contributes to the overall equilibrium constant of a sequence and therefore can determine the direction of conversion that is thermodynamically favorable. Any conversion can be made favorable by coupling to an appropriate number of ATP-to-ADP conversions.
7. Metabolic sequences occur in oppositely directed pairs, which are controlled by regulatory enzymes.
8. Regulatory enzymes respond to signals in such a way that rates of biosynthesis are controlled by the need for product.
9. Metabolism is regulated primarily by adjustment of the ratios by which intermediates are partitioned at metabolic branchpoints.
10. Different procedures are used for analysis of simple and complex pathways. Analysis of single-step pathways often begins with the isolation and characterization of the enzyme involved. During enzyme isolation each purification step is monitored by a specific assay that measures the conversion of substrate to product. Multistep pathway analysis ideally begins with complementation analysis, a genetic technique that entails the isolation of mutants with genetic blocks in each step of the pathways. Once the numbers of enzymes and intermediates are established, each enzyme can be isolated with the help of a specific assay.
11. Radiolabeled compounds are most useful for pathway analysis in two respects: first, they permit detection of pathway intermediates with great sensitivity, and second, they can be used as tracers for determining the order of intermediates in a pathway.
12. Inhibitors that block specific steps in a pathway play the same role as genetic blocks in the *in vitro* analysis of a pathway.
13. A complete understanding of a pathway requires parallel investigations *in vitro* and *in vivo*. *In vitro* analyses permit a detailed study of isolated components. *In vivo* observations substantiate the biologic significance of *in vitro* observations.

Problems

1. Consider the following relationships among the four major classes of biological molecules. What similarities and differences can you see in the chemical relationships between the right and left columns?

Small Molecule	Large Molecule
Nucleotide	Nucleic acid
Amino acid	Protein
Monosaccharide	Polysaccharide
Fatty acid	Lipid

2. What is a metabolic pathway?

3. What is the relationship between catabolism and anabolism?

4. Besides a difference in the number of phosphate moieties, what are the major biochemical differences between NAD^+ and $NADP^+$?

5. What is the primary advantage of subcellular compartments to intermediary metabolism?

6. Using only figure 11.2 determine if the conversion of glucose to pyruvic acid is an oxidation or a reduction or neither.

7. Explain why each of the following statements is false in terms of efficient metabolic regulation:
   a. Most enzymes operate *in vivo* near V_{max},
   b. End-product inhibition usually occurs at the last or next-to-last enzyme in a metabolic pathway.
   c. Catabolic pathways tend to diverge from a single metabolite.
   d. The enzymes regulated in a metabolic pathway usually exhibit simple Michaelis-Menten kinetics.
   e. Energy charge is unimportant in the regulation of anabolic sequences but is of primary importance in the regulation of catabolic sequences.
   f. Enzymes that catalyze a sequence of reactions are rarely grouped in multienzyme complexes.
   g. Compartmentalization of metabolic pathways is seldom a regulatory strategy the cell uses.

8. How is it possible that both the glycolytic degradation of glucose to lactate and the reverse process, formation of glucose from lactate (gluconeogenesis), are energetically favorable?

9. What is the metabolic advantage of having the "committed step" of a pathway under strict regulation?

10. Theoretically, the reactions shown below constitute a futile cycle. Explain.

$$\text{Glucose} + \text{ATP} \xrightarrow{\text{Glucokinase}} \text{Glc-6-P} + \text{ADP}$$

$$\text{Glc-6-P} + \text{HOH} \xrightarrow{\text{Glucose-6-phosphatase}} \text{Glucose} + \text{phosphate}$$

In the liver cell, the enzymes are spatially separated: Glucokinase in the cytosol and glucose-6-phosphatase in the endoplasmic reticulum. Does this separation influence the futile cycle?

Solutions

1. The transition from left to right involves dehydration reactions, while from right to left involves hydrolysis reactions. The items on the left can be thought of as monomers while only the first three items on the right are polymers. A typical lipid contains fatty acids but is not considered to be a polymer.
3. Catabolism involves pathways composed of enzymes and chemical intermediates that are primarily involved in the breakdown of large molecules into small molecules, often by oxidation processes. Anabolism is the collection of the enzymes and chemical intermediates involved in the biological synthesis of larger molecules from smaller molecules. Anabolism often involves reduction processes. Collectively anabolism and catabolism are called metabolism.
5. The advantage of subcellular compartments is that a specific pathway can be isolated from comparable pathways that might use similar or identical chemical intermediates. This simplifies regulation of the various processes.
7. a. The concentration of most metabolites measured under steady-state conditions in the cell usually does not exceed the K_m value. For an enzyme whose reaction can be described by simple Michaelis-Menten kinetics, the observed velocity, v, is 0.5 V_{max} if substrate concentration equals the K_{max} value. The velocity of most enzymes is likely significantly less than V_{max} *in vivo.*
   b. End-product inhibition usually occurs at the committed step in a metabolic pathway or at a branch point in the pathway. Regulation of the enzyme(s) at these steps in the metabolic pathway prevents the accumulation of intermediates in the pathway when the cell has limited demand for the end product. Regulation by individual metabolites at a branch point of the pathway inhibits their own production without simultaneously inhibiting the production of product from the other branch. Each product may cumulatively inhibit an enzyme catalyzing a precursor common to each end product.
   c. Catabolic pathways tend to be convergent rather than divergent. Metabolic convergence of precursors into common intermediates of a metabolic pathway provides an efficient route for the metabolism of a variety of metabolites by a limited number of enzymes. For example, the catabolism of glucose, fructose, and glycerol to the common intermediate glyceraldehyde-3-phosphate in the liver illustrates the convergence of the glycolytic pathway. Anabolic pathways tend to be divergent, with the synthesis of several end products from a common precursor.
   d. Enzymes that are regulated in metabolic pathways most frequently exhibit cooperative kinetic responses rather than hyperbolic response with respect to substrate concentration and are frequently responsive to allosteric regulation by products, energy charge, or concentration ratio of NAD(P)H/NAD(P)*. The rate of enzymatic activity over a narrow range of substrate concentration can be changed dramatically by allosteric activators or inhibitors. Such dramatic changes in velocity in response to small changes in substrate concentration are not observed with enzymes exhibiting Michaelis-Menten kinetics.
   e. Energy charge is a means of expressing the fraction of adenylate nucleotides that are high free energy compounds: ATP and ADP. The expression

$$E.C. = \frac{[ATP] + [ADP]/2}{[ATP] + [ADP] + [AMP]}$$

where E.C. represents the energy charge varying theoretically between 1 (all ATP) to 0 (all AMP). The energy charge is a metabolic signal for both anabolic (biosynthetic) and catabolic (degradative) metabolic pathways. Low energy charge signals the cell that a need for ATP formation exists, and pathways (glycolysis and Krebs cycle) leading to ATP formation are activated. Anabolic pathways that demand high ATP concentrations are inhibited at low energy charge. In the latter case, the ATP required to drive biosynthesis is in low supply. Conversely, increased energy charge inhibits pathways leading to ATP formation and activates anabolic pathways.

f. Formation of multienzyme complexes is a strategy frequently used for efficient catalysis and control of metabolic pathways. Substrates diffuse shorter distances between active sites in multienzyme complexes than if the enzymes were not organized. Frequently, intermediates covalently bound to cofactors (e.g., lipoamide, biocytin) are moved among active sites within the complex, effectively trapping intermediates in the complex and increasing the concentration of substrates at the enzyme active site.

g. Separation of catabolic and anabolic pathways diminishes the likelihood of futile cycling of metabolites. Enzymes catalyzing the β-oxidation of fatty acids are located in the mitochondrial matrix, whereas enzymes catalyzing the synthesis of palmitate are located in the cytosol.

9. The "committed step" in a reaction sequence steers the metabolite to a sequence of reactions whose intermediates usually have no other function in the cell. Control of the committed step prevents wasteful accumulation of these single-purpose intermediates and obviates the necessity of controlling each enzyme in a pathway. Enzymes that catalyze reactions just past a branch point in a pathway are likely inhibited by their respective products or the end product of the branch.

12 Glycolysis, Gluconeogenesis, and the Pentose Phosphate Pathway

Summary

In this chapter we discuss the structure and anaerobic metabolism of the carbohydrates involved in energy metabolism. We focused on the following points.

1. Carbohydrates include monosaccharides, oligosaccharides, and polysaccharides.
2. Monosaccharides tend to form ring structures known as intramolecular hemiacetals (or hemiketals), especially when the product is a five-member (furanose) ring or a six-member (pyranose) ring. Depending on which way the ring forms about the reactive carbon, the structure is called an α- or β-hemiacetal.
3. Monosaccharides are linked together by glycosidic bonds to form oligosaccharides and polysaccharides.
4. Polysaccharides of glucose function in two distinct roles. Some serve for storage of chemical energy, and others fill structural roles. The structural polysaccharides like cellulose have a β (1,4) linkage between monomeric residues, whereas the polysaccharides involved in the storage of chemical energy have an α (1,4) linkage between adjacent monomeric residues.
5. Glycolysis involves the breakdown of energy-storage polysaccharides or glucose to the 3-carbon acid, pyruvate. Its function is to produce energy in the form of ATP and 3-carbon intermediates for further metabolism. Glycolysis occurs under both aerobic and anaerobic conditions.
6. The breakdown of storage polysaccharides yields glucose-1-phosphate, but the first step in the metabolism of free glucose is phosphorylation to glucose-6-phosphate. These two compounds are freely interconvertible with fructose-6-phosphate.
7. The three sugar phosphates just named make up the hexose phosphate pool, which is a major crossroad of energy metabolism. Glucose-1-phosphate is produced from storage polysaccharides and is used in their synthesis; glucose-6-phosphate is produced from glucose and is the precursor for blood glucose in mammals and the starting material for the pentose phosphate pathway; fructose-6-phosphate is produced by gluconeogenesis, the reverse of glycolysis, in which 6-carbon sugars are synthesized from pyruvate.
8. Glycolysis and gluconeogenesis proceed through three pools of intermediates that are close to equilibrium. The pools are the hexose monophosphate pool, the aldolase pool, and the three-carbon pool.
9. Within each pool, the direction of reaction is determined by simple mass-action considerations. The direction of overall conversion (glycolysis or gluconeogenesis) depends on the rates at which the connecting reactions supply and remove materials from the pool.
10. Two out of three of the connecting reactions between pools have large free energies of reaction and are far from equilibrium in the metabolizing cell. They are subject to regulatory signals that indicate the metabolic needs of the cell or organism. As a result of those regulatory interactions, polysaccharide is stored when nutrients are available in excess of current needs, and storage compounds are broken down to supply energy when necessary.
11. In mammals, the liver is a special organ. It is a major site of storage of glycogen, and it regulates the level of glucose in the blood. When the blood glucose level is low, the liver replenishes it by hydrolysis of glucose-6-phosphate, which is made simultaneously by breakdown of glycogen and by gluconeogenesis. When the liver's supply of glycogen is depleted, blood glucose is

replenished exclusively by gluconeogenesis. Hormonal signals trigger the liver to respond to the need for maintaining the blood glucose level.

12. The pentose phosphate pathway serves a variety of functions: (1) the production of NADPH for biosynthesis; (2) the production of ribose, required mainly for nucleic acid synthesis, and (3) the interconversion of a variety of phosphorylated sugars.

Problems

1. Why are there more D-aldohexoses (fig.12.2) than D-ketohexoses (fig. 12.3)?

2. Draw the structures of α-sophorose, α-melibiose, and lactulose. Sophorose is two glucoses linked β (1-2). Melibiose is a galactose linked to a glucose by an α (1-6) bond. Lactulose is galactose linked β (1-4) to fructose.

3. A tool that has often been used to deduce ring sizes and linkages between sugars is methylation with methyl iodide. The methylated sugar is then hydrolyzed and the location of the methyl groups determined. A disaccharide called trehalose has been obtained from a number of vegetable sources. Trehalose is composed only of glucose, and when methylated and hydrolyzed, only 2,3,4,6-tetramethyl-glucose results. Using the same procedure, maltose (fig. 12.8) gives 2,3,4,6-tetramethylglucose and 2,3,6-trimethylglucose in equal molar ratios. Propose a structure(s) for trehalose.

4. The substance glucuronic acid can be considered to be glucose with carbon 6 oxidized to the carboxylic acid level. Glucuronic acid β-glycosides are the metabolic fate of a number of obnoxious compounds found in plants. Draw the structure of glucuronic acid. From its name can you guess where it was first found?

5. Aldoses and ketoses that have the potential to form open chains (i.e., have hemiacetal or hemiketal groups) are referred to as reducing sugars. This name is derived from the fact they can readily reduce Ag^{+} to Ag or Cu^{2+} to Cu^{1+}. Examine the disaccharides in figure 12.8 and determine which substances are reducing sugars. Is trehalose (problem 3) a reducing sugar?

6. The polysaccharide inulin is an energy-storage substance found in the tubers of a number of plants. The major part of inulin is composed of fructoses linked β(2-1). Draw the structure of inulin.

7. Which glucose carbons (use glucose carbon numbers) are lost as CO_2 in the conversion of pyruvate into ethanol (fig. 12.13).

8. In the conversion of pyruvate to phosphoenolpyruvate in gluconeogenesis (fig. 12.26) an addition of CO_2 is followed by a decarboxylation. Why would nature add an item only to remove it in the next step? Is the carbon added the same as the one removed?

9. Glucose can be purchased with essentially any specific carbon labeled with ^{14}C. If your objective is to assess the relative importance of glycolysis versus the pentose phosphate pathway in a particular tissue, how might you use radioactive glucose to answer this question?

10. After reading about glycolysis and gluconeogenesis do you find anything unusual about the name pyruvate kinase?

11. An intermediate in the interconversion of glycerate-3-phosphate to glycerate-2-phosphate is glycerate-2,3-bisphosphate. Where have we encountered this bisphosphate before?

12. The common yeast genus *Rhodotorula* is thought to be missing the enzyme phosphofructokinase yet seems to survive nicely. Can you propose a pathway around phosphofructokinase?

13. Why are the mechanisms of the enzymes that interconvert glucose-6-phosphate and fructose-6-phosphate and the enzyme that interconverts dihydroxyacetone phosphate and glyceraldehyde-3-phosphate virtually identical? Compare the mechanisms of phosphoglycerate mutase and phosphoglucomutase. Why are they so similar?

14. Suppose that you have isolated a facultative microorganism that you are growing anaerobically in a medium containing a carbohydrate. Explain, on the basis of your knowledge of metabolism, why each of the following statements about the fermentation is false:
    a. The culture must be growing on glucose because bacteria ferment few other compounds.
    b. The products of the fermentation must be more highly oxidized than the substrates, otherwise no energy is conserved.
    c. The culture cannot be producing any CO_2.

15. A yeast culture is fermenting glucose to ethanol. To ensure that the CO_2 released during fermentation is radiolabeled, what carbon(s) of glucose must be labeled with ^{14}C?

16. Assume that a mutant form of glyceraldehyde-3-phosphate dehydrogenase was found to hydrolyze the oxidized enzyme-bound intermediate with water rather than phosphate.
    a. Write a chemical reaction that describes the hydrolysis, showing the structures of the products.
    b. What would be the effect, if any, on the ATP yield from glycolysis of glucose to lactate?
    c. What would be the effect of the mutation on an obligate aerobic microorganism?

17. Suppose that you are seeking bacterial mutants with altered triose phosphate isomerase (TPI). The organism of interest is known to use the glycolytic pathway with the production of lactate.
    a. Explain why the absence of TPI would be lethal to an organism fermenting glucose exclusively through the glycolytic pathway.
    b. Suppose that you have an organism that uses glycolysis and an oxidative pathway as energy sources. Mutants of that organism having only 10% of the TPI activity present in the wild-type cells grew slowly on glucose under anaerobiosis but grew faster aerobically. Explain the metabolic basis of the observation.
    c. You constructed a plasmid that would direct the synthesis of dihydroxyacetone phosphate phosphatase and introduced the plasmid into the mutant organism described in part (b). Predict whether the plasmid-bearing organism with an active DHAP phosphatase could grow on glucose or glycerol either anaerobically or aerobically. What are the metabolic considerations you used to make your predictions?

18. There are two sites of ADP phosphorylation in glycolysis. These processes are called substrate level phosphorylations. Arsenate (AsO_4^{3-}), an analog of phosphate, uncouples ATP formation resulting from glyceraldehyde-3-phosphate oxidation but not that resulting from dehydration of glycerate-2-phosphate. Explain.

19. The disaccharide sucrose can be cleaved by either of two methods:

$$\text{Sucrose} + H_2O \xrightarrow{\text{Invertase}} \text{glucose} + \text{fructose}$$

$$\text{Sucrose} + P_i \xrightarrow{\text{Sucrose phosphorylase}} \text{glucose-1-phosphate} + \text{fructose}$$

    a. Given that the $\Delta G^{\circ\prime}$ value for the invertase reaction is –7.0 kcal/mole, calculate the $\Delta G^{\circ\prime}$ value for the sucrose phosphorylase catalyzed reaction. Assume that the $\Delta G^{\circ\prime}$ for hydrolysis of glucose-1-phosphate is –5 kcal/mole. Based on the calculated value of $\Delta G^{\circ\prime}$, calculate the equilibrium constant for sucrose phosphorylase at 25°C.

b. Explain the metabolic advantage to the cell of cleaving sucrose with phosphorylase rather than with invertase.

20. What concentration of glucose would be in equilibrium with 1.0 mM glucose-6-phosphate, assuming that hexokinase is present and the concentration ratio of ATP to ADP is 5? Is it reasonable to expect an actively metabolizing cell to maintain the concentration of glucose necessary to sustain the concentration of glucose-6-phosphate at 1 mM?

21. 2-Phosphoglycerate and phosphoenolpyruvate differ only by dehydration between C-2 and C-3, yet the difference in the $\Delta G^{\circ\prime}$ of hydrolysis is about –12 kcal/mole. How does dehydration "trap" so much chemical energy?

22. Elevated pyruvate concentration inhibits the heart muscle lactate dehydrogenase (LDH) isoenzyme but not the skeletal muscle LDH isoenzyme.
   a. What would be the consequence of having only the heart isoenzyme form in skeletal muscle?
   b. Would there necessarily be a negative consequence if the skeletal muscle isoenzyme were the only LDH in heart muscle?

23. We stated that the equilibrium constant for the pyruvate kinase reaction is 10^6. Assume that the steady-state concentration of ATP is 2 mM and of ADP is 0.2 mM. Calculate the concentration ratio of pyruvate to PEP under these conditions. Does your calculation support or refute the assertion that the pyruvate kinase reaction is metabolically irreversible?

24. In gluconeogenesis, the thermodynamic barrier imposed by pyruvate kinase is overcome by coupling two separate reactions for the synthesis of PEP from pyruvate.
   a. Write the two chemical reactions used to bypass the pyruvate kinase reaction.

b. Calculate the overall $\Delta G^{\circ\prime}$ of the two reactions you wrote in part (a). (Assume that GTP is the thermodynamic equivalent of ATP.) What can you now surmise about the feasibility of PEP formation from pyruvate by this route?

25. Write a chemical reaction for the $NADP^+$-dependent oxidation of 6-phosphogluconate to ribulose-5-phosphate.

26. Fructose-2,6-bisphosphate is a potent activator of the liver phosphofructokinase (PFK-1) and a potent inhibitor of liver fructose-1,6-bisphosphate phosphatase (FBPase-1). Fructose-2,6-bisphosphate is the product of a second phosphofructokinase (PFK-2) and is hydrolyzed to fructose-6-phosphate by FBPase-2. The activities of PKF-2 and FBPase-2 reside on a single, bifunctional protein in liver. The bifunctional protein is under glucagon control imposed via cAMP. (H.-G Hers, *Arch. Biol Med. Exp.* 18:243–251, 1985.)
   a. Under what metabolic conditions is PKF-2 active? FBPase-2?
   b. Gluconeogenesis in liver is stimulated by the hormone glucagon. The activity of PFK-2/FBPase-2 bifunctional enzyme under glucagon regulation shifts from an active PFK-2 to an inactive PFK-2. Inactivation of the PFK-2 alone would still not be adequate to stimulate gluconeogenesis sufficiently for the organism. Explain.
   c. Cyclic AMP-dependent phosphorylation of PFK-2/FBPase-2 not only inhibits PFK-2 but stimulates FBPase-2. Under these conditions, gluconeogenesis is sufficiently rapid to meet cellular demand. Explain.
   d. What would you predict as the relative activities of the following enzymes in the liver of a rat made diabetic through chemical means (administration of alloxan or streptozotocin): PFK-2, FBPase-2, PFK-1, FBPase-1, pyruvate carboxylase, PEP carboxykinase?

27. Muscle pyruvate kinase (PK) responds hyperbolically to its substrate, PEP, but the liver form of the enzyme responds sigmoidally. Fructose-1.6-bisphosphate is an allosteric activator of liver pyruvate kinase, but it apparently has no effect on the muscle enzyme.
   a. If liver PK responded hyperbolically to PEP and were otherwise unregulated, how might gluconeogenesis be affected?
   b. What is the metabolic advantage of having the liver PFK activated by fructose-1.6-bisphosphate?

Solutions

1. The number of sugars is 2^n where n is the number of chiral centers but does not include the chiral carbon involved in producing the D-series of sugars. Because carbon two of ketoses is a carbonyl group, only the geometry of carbons 3 and 4 remain to produce the four D-ketohexoses ($2^2 = 4$). The carbonyl group on carbon 1 of the aldoses allow carbons 2, 3, and 4 to produce eight D-aldohexoses ($2^3 = 8$).
3. Trehalose (A) is two α-D-glucoses linked through carbon 1 of each sugar. However, the correct answer to this question would also include (B) two β-D-glucoses linked through carbon 1 and (C) one α-D-glucose linked through carbon 1 to carbon 1 of a β-D-glucose.

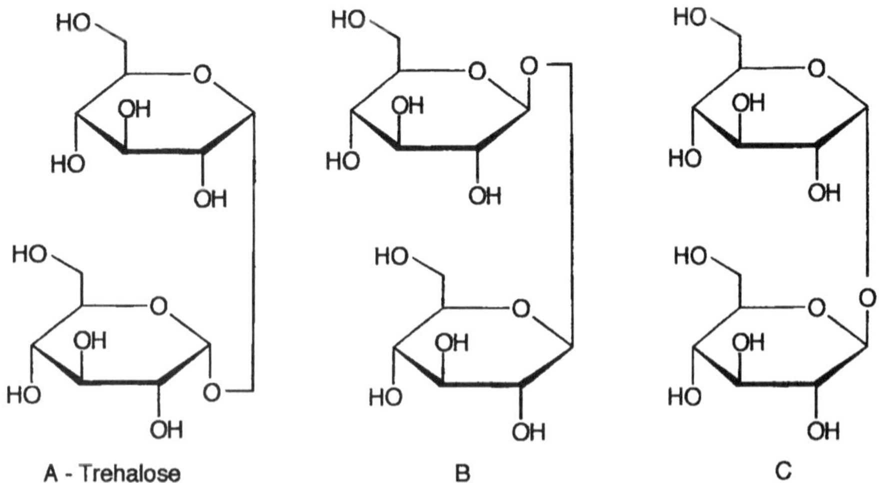

5. Maltose, lactose, and cellobiose are reducing sugars; sucrose and trehalose are non-reducing sugars.
7. Glucose carbons 3 and 4 are lost as CO_2 during ethanol production.
9. Carbon 1 of glucose is lost as CO_2 in an early reaction in the pentose phosphate pathway. Carbons 3 and 4 of glucose are lost as CO_2 with the formation of acetyl-CoA after glycolysis. Carbons 1, 2, 5, and 6 are lost in the various turns of the citric acid cycle. By assessing the rate of $^{14}CO_2$ production from ^{14}C-1-glucose versus ^{14}C-6-glucose, the relative significance of the different pathways in a tissue can be determined.
11. Glycerate-2,3-bisphosphate was encountered as an allosteric effector of hemoglobin (see page 103 in the text).
13. In the first example, the actual chemistry is identical: an aldose-ketose interconversion. Notice that the geometry of the alcohol on carbon 2 is identical. The second example has essentially the same chemistry. It is logical that nature could have stumbled upon the same chemical mechanism twice or more likely that the mechanism evolved once, and following gene duplication, the segments of the gene controlling the substrate specificity changed on one copy of the gene.
15. The carboxyl group of pyruvate is lost as CO_2 during the pyruvate decarboxylase-catalyzed formation of acetaldehyde. The pyruvate carboxyl group is formed by the oxidation of glyceraldehyde-3-phosphate. The aldehyde (C-1) carbon is derived directly from the aldolase-dependent cleavage of the frutose-1, 6-bisphosphate. In this cleavage, C-4 of glucose becomes C-1 of glyceraldehyde-3-P_i while C-3 of glucose becomes C-3 of dihydroxyacetone phosphate. DHAP is isomerized to Ga3P_i. In this isomerization, C-3 of DHAP (originally C-3 of glucose)

becomes C-1 of Ga3P. Thus, labeling either C-3 or C-4 of glucose will ensure that label is released as CO_2 upon fermentation to ethanol. The labeling scheme shown here indicates the carbons of fructose-1,6-bisphosphate, DHAP, $Ga3P_i$ and ethanol numbered to correspond to the corresponding carbon of glucose.

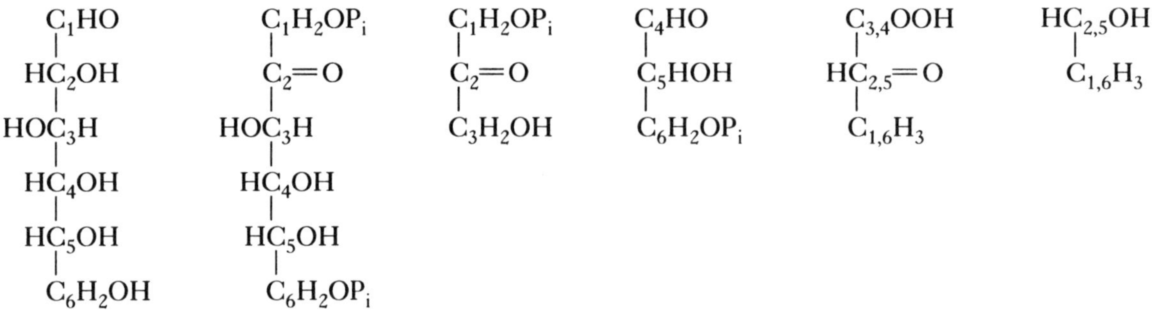

Glc	**FBP**	**DHAP**	**$Ga3P_i$**	**pyruvate**	**ethanol**
C_1HO	$C_1H_2OP_i$	$C_1H_2OP_i$	C_4HO	$C_{3,4}OOH$	$HC_{2,5}OH$
HC_2OH	$C_2{=}O$	$C_2{=}O$	C_5HOH	$HC_{2,5}{=}O$	$C_{1,6}H_3$
HOC_3H	HOC_3H	C_3H_2OH	$C_6H_2OP_i$	$C_{1,6}H_3$	
HC_4OH	HC_4OH				
HC_5OH	HC_5OH				
C_6H_2OH	$C_6H_2OP_i$				

17. a. Triose phosphate isomerase deficiency would inhibit conversion of DHAP to Ga3P and would cause accumulation of DHAP, preventing half of the glucose molecule (C1-C3) from being metabolized through the remainder of the glycolytic pathway. There would be a recovery of only 2 of the possible 4 moles of ATP from glucose, resulting in no net formation of ATP. In addition, DHAP, a product of the aldolase reaction, would likely reverse the aldolase (reaction) and eventually inhibit glycolysis. Either result would be lethal to a cell whose only energy source was glycolysis.

    b. The small amount of TPI activity would likely allow glycolysis to proceed slowly, but low energy (ATP) level will limit the growth rate under anaerobiosis. However, the yield of ATP is significantly greater when the pyruvate, formed during glycolysis, is oxidized to CO_2 and H_2O. Hence, the growth rate of the mutant should be correspondingly greater under aerobic growth conditions but not as great as the wild type.

    c. Cells expressing DHAP phosphatase would likely not grow anaerobically if glycolysis of glucose to lactate were the only pathway for ATP formation. The combined activities of TPI and DHAP phosphatase would be predicted to deplete the pool of triosephosphate and the yield of ATP per glucose would likely be less than 1.

    The cells might grow aerobically, depending on the competition among Ga3PDH, TPI, and DHAP phosphatase for the triosephosphate pool.

    Glycerol can be phosphorylated to α-glycerolphosphate and oxidized to DHAP. The organism expressing DHAP phosphatase would likely not grow anaerobically on glycerol, but might grow aerobically for the reasons described above.

19. a. Consider the reactions:

$$H_2O + \text{Sucrose} \rightarrow \text{fructose} + \text{glucose}\ \Delta G^{\circ\prime} = -7\ \text{kcal/mole}$$

$$\text{Glucose} + P_i \rightarrow \text{Glucose-1-}P_i + H_2O\ \Delta G^{\circ\prime} = +5\ \text{kcal/mole}$$

$$\text{Sucrose} + P_i \rightarrow \text{Glucose-1-}P_i + \text{fructose}\ \Delta G^{\circ\prime} = -2\ \text{kcal/mole}$$

$\Delta G^{\circ\prime} = -2.3RT \log K'_{eq}$ where 2.3RT is 1.36 kcal/mole at 25°C.

$$-2\ \text{kcal/mole} = -1.36\ \text{kcal/mole}\ \log K'_{eq}$$

$$\text{Log}\ K'_{eq} = 1.47$$

$$K'_{eq} = 30$$

b. Hexoses brought into the glycolytic pathway must be phosphorylated to provide the appropriate substrate for the glycolytic enzymes and to trap the sugar within the cell. Phosphorylation at the expense of ATP or group translocation at the expense of PEP are common methods to activate the sugar molecules. Sucrose phosphorylase uses the exergonic lysis of the glycosidic bond between the hemiacetal OH group of glucose and the hemiketal OH group of fructose to drive the endergonic phosphorylation of the hemiacetal C-1 OH group of glucose. Transfer of the phosphate to C-6, catalyzed by phosphoglucomutase, provides substrate for entry into the glycolytic pathway without addition of ATP. The net ATP yield will therefore be 3 rather than 2 moles of ATP per mole of glucose derived from sucrose. The fructose can be phosphorylated by ATP and used in the glycolytic pathway.

21. The free energy available from a reaction depends on the energies of the products compared with the substrates. Dehydration of 2-phosphoglycerate "traps" phospho-(enol)pyruvate in the enolate form. The hydrolysis products of PEP are phosphate and the enol form of pyruvate, but (enol) pyruvate is significantly less stable that (keto) pyruvate and rapidly tautomerizes to the more stable keto form. The tautomerization drives the reaction strongly toward products, resulting in a larger free energy difference between substrate and product.

23. Given the ratio of ATP/ADP and the K'_{eq} of 10^6, the equilibrium ratio of [Pyr]/[PEP] would be about 10^5. This calculation supports the metabolic irreversibility of the pyruvate kinase reaction.

25.

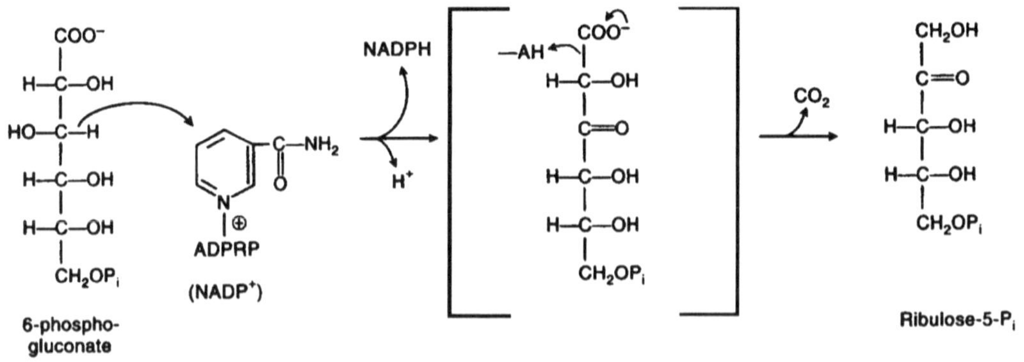

6-phospho-gluconate + ADPRP ($NADP^+$) → NADPH + H^+ → [3-keto intermediate] → CO_2 + Ribulose-5-P_i

27. a. Unregulated hepatic PK theoretically could become part of a futile cycle.

$$\text{PEP} + \text{ADP} \rightarrow \text{Pyr} + \text{ATP}$$

$$\text{Pyr} + CO_2 + \text{ATP} \rightarrow \text{oxaloacetate}_{mito} + \text{ADP} + P_i$$

$$\text{OAA}_{mito} + \text{NADH} \rightarrow \text{L-malate}_{mito} + \text{NAD}^+$$

$$\text{L-malate}_{mito} \rightarrow \text{L-malate}_{cyto}$$

$$\text{L-malate}_{cyto} + \text{NAD}^+ \rightarrow \text{OAA}_{cyto} + \text{NADH}$$

$$\text{OAA}_{cyto} + \text{GTP} \rightarrow \text{PEP} + \text{GDP} + CO_2$$

Net: GTP → GDP + P_i plus formation of cytosolic NADH at the expense of mitochondrial NADH.

b. Activation by fructose-1,6-bisphosphate decreases $S_{0.5}$ for PEP, and increases PK activity at a given PEP concentration. During gluconeogenesis, the fructose-1,6-bisphosphate concentration should diminish due to the hydrolytic activity of the FBPase-1. The low FBP concentration, coupled with the elevated ATP levels, could inhibit the hepatic pyruvate kinase.

13 The Tricarboxylic Acid Cycle

Summary

This chapter is mainly concerned with the contribution of the tricarboxylic acid cycle to carbohydrate metabolism. The TCA cycle is the main source of electrons for oxidative phosphorylation, and thereby the major energetic sequence in the metabolism of aerobic cells or organisms. It serves as the main distribution center of metabolism, receiving carbon from the degradation of carbohydrates, fats, and proteins and, when it is appropriate, supplying carbon compounds for the synthesis of carbohydrates, fats, or proteins. Every aspect of the metabolism of an aerobic organism is directly dependent on the TCA cycle.

1. The TCA cycle begins with acetyl-CoA, which is obtained either by oxidative decarboxylation of pyruvate available from glycolysis or by oxidative cleavage of fatty acids.
2. The acetyl-CoA transfers its acetyl group to oxaloacetate, thereby generating citrate. In a cyclic series of reactions, the citrate is subjected to two successive decarboxylations and four oxidative events, leaving a four-carbon compound malate from which the starting oxaloacetate is regenerated.
3. Only a single ATP is directly generated by a turn of the TCA cycle. Most of the energy produced by the cycle is stored in the form of reduced coenzyme molecules, NADH and $FADH_2$. Reoxidation of these compounds (see chapter 14) liberates a large amount of free energy, which is captured in the form of ATP.
4. Some of the main biosynthetic pathways begin with intermediates in the TCA cycle. When intermediates in the cycle are used as starting materials for biosynthesis, they must be replenished to keep the cycle operating. When carbohydrates are being metabolized, TCA cycle intermediates are replenished by production of oxaloacetate from pyruvate.
5. The glyoxylate cycle permits growth on a two-carbon source. The glyoxylate cycle bypasses the two steps of the TCA cycle in which CO_2 is released. Furthermore, two molecules of acetyl-CoA are taken in per turn of the cycle rather than just one, as in the TCA cycle. The net result is the conversion of two molecules of two carbons each into one four-carbon compound, succinate. Two additional enzymes are needed for operation of the glyoxylate cycle: isocitrate lyase and malate synthase.
6. Degradation of amino acids produces a number of intermediates, among which are α-ketoglutarate, succinyl-CoA, and oxaloacetate. α-Ketoglutarate and succinyl-CoA can be oxidized to oxaloacetate, but the cycle as such cannot oxidize oxaloacetate further. Oxaloacetate is oxidized further by first converting it to phosphoenolpyruvate. This permits the total oxidation of oxaloacetate to CO_2 by the enzymes of the TCA cycle.
7. The TCA cycle is regulated to ensure that its level of activity corresponds closely to cellular needs. In its primary role as a means of oxidizing acetyl groups to CO_2 and water, the TCA cycle is sensitive both to the availability of its substrate, acetyl-CoA, and to the accumulated levels of its principal end products, NADH and ATP. Other regulatory parameters to which the TCA cycle is sensitive include NAD^+, ADP, acetyl-CoA, succinyl-CoA, and citrate. The major known sites for regulation of the cycle include two enzymes outside the cycle (pyruvate dehydrogenase and pyruvate carboxylase) and three enzymes inside the cycle (citrate synthase, isocitrate dehydrogenase, and α-ketoglutarate dehydrogenase). All of these sites of regulation represent important metabolic branchpoints.

Problems

1. In the conversion of isocitrate to α-ketoglutarate, oxidation and decarboxylation steps occur. Figure 13.9 shows the oxidation step first. Can you suggest a reason why the oxidation step is first?

2. In the late 1930s the tricarboxylic acid cycle (fig. 13.3) contained *cis*-aconitate, but more modern versions of the cycle (since the late 1950s) do not. The involvement of *cis*-aconitate is still shown in many nonbiochemistry texts. Can you explain why *cis*-aconitate is deleted from modern versions of the tricarboxylic acid cycle?

3. Notice in the initial steps of the tricarboxylic acid cycle, citrate is converted to isocitrate, which is then oxidized. Why didn't nature "just" oxidize citrate and save an enzymatic step?

4. Ethylene glycol (ethane-1,2-diol), a major component of antifreeze, is readily metabolized by many organisms that have the glyoxylate cycle. Can you propose a sequence of catabolic steps that can explain the catabolism of ethylene glycol?

5. Many organisms can live on glutamate as their sole carbon and nitrogen source. Assuming glutamate is converted into α-ketoglutarate, produce a scheme that completely oxidizes glutamate.

6. Notice the intermediate in the reaction of citrate synthase (fig. 13.7). Do you think at some time in the future evolution will produce a variety of citrate synthase that recovers the energy in the thioester, analogous to the production of GTP (ATP) by succinyl-CoA synthase (page 291)? Would this energy recovery have any effect on the thermodynamics of the tricarboxylic acid cycle?

7. Summarize in the simplest words the portion of the tricarboxylic acid cycle that is bypassed by the glyoxylate cycle?

8. The reactions catalyzed by isocitrate dehydrogenase and α-ketoglutarate dehydrogenase are both oxidative decarboxylation reactions. How similar are the reactions?

9. What effect would the following have on the control of the citric acid cycle: (a) a sudden influx of acetyl-CoA from the degradation of fatty acids, and (b) a sudden need for heme biosynthesis?

10. Do all of the metabolites associated with the tricarboxylic acid cycle have free access across the mitochondrial membrane?

11. Using your knowledge of metabolism, determine whether the following statements are true or false and explain the reasoning behind your decision.
    a. Dihydrolipoamide dehydrogenase catalyzes the only oxidation-reduction reaction in the pyruvate dehydrogenase complex.
    b. Hydrolysis of the thioester bond of acetyl-CoA yields insufficient energy to drive phosphorylation of ADP.
    c. The methyl group of each acetyl-CoA molecule entering the TCA cycle is derived from the methyl group of pyruvate.

d. Even if aconitase were unable to discriminate between the two ends of the citrate molecule, the CO_2 released would still come from the oxaloacetate rather than the acetyl-CoA substrate of the citrate synthase reaction.
e. Malate cannot be converted to fumarate because the TCA cycle is unidirectional.

12. Assume that you have a buffered solution containing pyruvate dehydrogenase and all the enzymes of the TCA cycle but none of the cycle intermediates.
    a. If you add 3 μmoles each of pyruvate, CoASH, NAD^+, GDP, and P_i, how much CO_2 evolves? What other products form?
    b. In addition to the reagents in (a), you add 3 μmoles each of the TCA cycle intermediates. How much CO_2 evolves? Explain.
    c. If you were to add an electron acceptor that reoxidized NADH to the system described in (a), would there be increased CO_2 evolution? Why or why not?
    d. Explain the effect on CO_2 evolution of adding the NADH-reoxidizing system to the system described in (b), assuming that you also added excess GDP and P_i?

13. What would you expect to be the metabolic consequences of the following mutations in yeast?
    a. Inability to synthesize malate synthase.
    b. Pyruvate carboxylase that is not activated by acetyl-CoA.
    c. Pyruvate dehydrogenase that is inhibited by acetyl-CoA more strongly than is the wild-type enzyme.

14. The substrate hydroxypyruvate

$$HO-CH_2-\overset{\overset{\large O}{\|}}{C}-\overset{\overset{\large O}{\|}}{C}-O^-$$

Hydroxypyruvate

is metabolized to pyruvate in a five-step process requiring the four intermediates whose structures are shown on the following page.

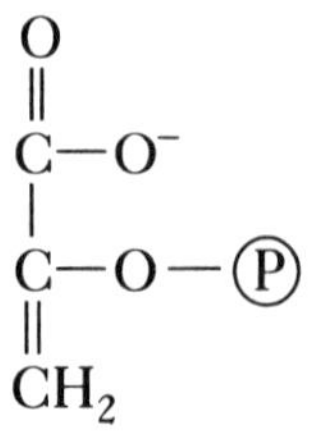

$$\begin{array}{c} O \\ \| \\ C{-}O^- \\ | \\ C{-}O{-}Ⓟ \\ \| \\ CH_2 \end{array}$$

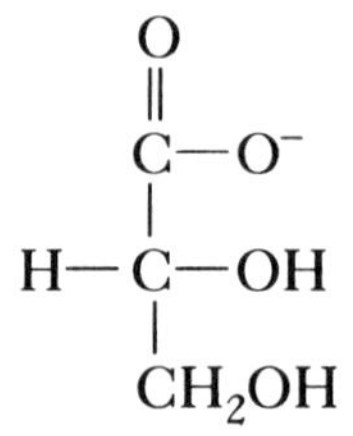

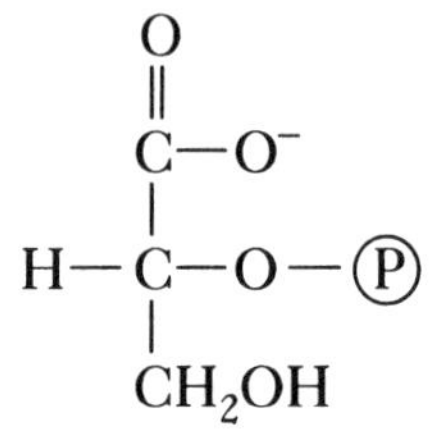

A

$$\begin{array}{c} O \\ \| \\ C{-}O^- \\ | \\ H{-}C{-}OH \\ | \\ CH_2OH \end{array}$$

B

$$\begin{array}{c} O \\ \| \\ C{-}O^- \\ | \\ H{-}C{-}O{-}Ⓟ \\ | \\ CH_2OH \end{array}$$

C

$$\begin{array}{c} O \\ \| \\ C{-}O^- \\ | \\ H{-}C{-}OH \\ | \\ CH_2{-}O{-}Ⓟ \end{array}$$

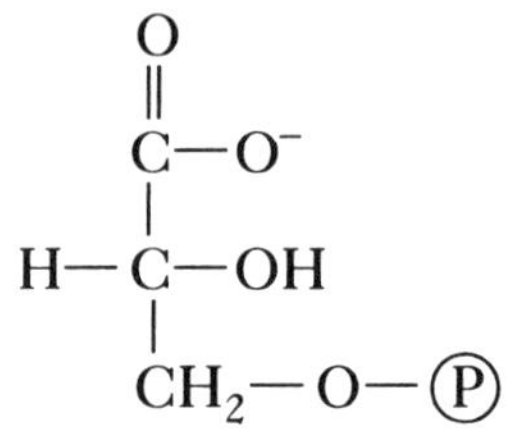

D

The letter designating the intermediate does not necessarily reflect the order in which it is used. The metabolic conversion requires NADH and catalytic quantities of both ATP and ADP. Assume that the pathway begins with an NADH-mediated reaction.

a. Designate the order in which the intermediates are used in the metabolism of hydroxypyruvate to pyruvate.
b. Write an overall equation for the metabolism of hydroxypyruvate to pyruvate.
c. Name each of the intermediates (A–D) and indicate which are intermediates in the glycolytic pathway.
d. Explain why only catalytic rather than stoichiometric amounts of ADP and ATP are required in the pathway.

15. Under anaerobic conditions, *E. coli* synthesizes an NADH-dependent fumarate reductase rather than succinate dehydrogenase, the flavoprotein that oxidizes succinate to fumarate.
    a. Write an equation for the reaction catalyzed by fumarate reductase.
    b. NADH produced by the glyceraldehyde-3-phosphate dehydrogenase reaction is reoxidized by reducing an organic intermediate. Rather than reduce pyruvate to lactate, anaerobic *E. coli* utilize the fumarate reductase. However, under anaerobiosis, the activity of α-ketoglutarate dehydrogenase is virtually nonexistent. Show how fumarate is formed, using reactions beginning with PEP and including the necessary TCA cycle enzymes. (Spiro, S., and J. R. Guest, *TIBS.* 16:310–314 (1991).)
    c. What is the metabolic advantage to anaerobic *E. coli* in using the fumarate reductase pathway rather than lactate dehydrogenase to reoxidize NADH?

16. Consider the glyceraldehyde-3-phosphate dehydrogenase-phosphoglycerokinase enzymes of glycolysis and the succinate thiokinase of the TCA cycle. Compare the mechanisms of incorporation of inorganic phosphate into the respective nucleoside diphosphates.

17. The pyruvate dehydrogenase complex may have been regulated by phosphorylation of any one of the three different enzymes in the complex, yet regulation occurs on the first enzyme of the complex. How is regulation of the complex consistent with the regulation observed in metabolic pathways whose enzymes are not physically associated?

18. Although there is no net synthesis of glucose from acetyl-CoA in mammals, acetyl-CoA has two major functions in gluconeogenesis. Explain the functions of acetyl-CoA in the synthesis of glucose from lactate in mammalian liver.

Solutions

1. The oxidation occurs first, creating a β-keto carboxylate which readily loses CO_2 via a resonance stabilized carbanion.
3. Citrate is a tertiary alcohol, which does not oxidize readily. Isocitrate, a secondary alcohol, readily oxidizes to a ketone (an intermediate in fig. 13.9).
5. The following sequence will do the task: α-ketoglutarate, the tricarboxylic acid cycle to oxaloacetate, to phosphoenolpyruvate, to pyruvate, to acetyl-CoA, into the tricarboxylic acid cycle.
7. Both of the oxidative-decarboxylation steps in the tricarboxylic acid cycle are by-passed by the glyoxylate cycle.
9. a. Increased concentrations of acetyl-CoA slows the activity of pyruvate dehydrogenase.
   b. The removal of succinyl-CoA for heme synthesis removes any control it has over lowering the activity of alpha ketoglutarate dehydrogenase and citrate synthetase.
11. a. False. Pyruvate dehydrogenase catalyzes reduction of the lipoamide disulfide concomitantly with oxidation and transfer of the hydroxyethyl group from thiamine pyrophosphate. The hydroxyethyl group is oxidized to form a thioester with one sulfhydryl group of lipoamide. The acetyl group is subsequently transferred to CoASH, forming the thioester adduct, acetyl-CoA. During the oxidation of the α-keto carbon, the disulfide (oxidized form) of lipoamide is

reduced to the dithiol form. Dihydrolipoamide dehydrogenase, a flavoprotein, catalyzes oxidation of dihydrolipoamide and reduction of NAD^+ to NADH.

b. False. Hydrolysis of acetyl-CoA thioester should yield as much free energy as succinyl-CoA hydrolysis. Succinate thiokinase catalyzes the substrate level phosphorylation of GDP at the expense of succinyl-CoA hydrolysis. However, in the TCA cycle, there is no enzymatic pathway to couple the hydrolysis of the acetyl-CoA to the activation of orthophosphate and subsequent transfer of the activated phosphate to ADP. Succinate thiokinase provides such an enzymatic pathway to couple the energy of succinyl-CoA hydrolysis to GDP.

c. False. The TCA cycle is a versatile pathway for the oxidation of various substrates, including carbohydrates, fatty acids, and lipids. Thus, the methyl group of acetyl-CoA could be derived from pyruvate, from β-oxidation of long chain fatty acids, or from amino acid catabolism.

d. False. The CO_2 molecule released by oxidative decarboxylation of isocitrate (ICDH) is derived from the carboxyl group of oxaloacetate with which the acetyl-CoA was condensed. However, the CO_2 released from α-ketoglutarate does depend on discrimination between the ends of the citrate molecule. If the aconitase reaction were random, half the CO_2 would arise from an oxaloacetate carboxyl group and half from the acetate carboxyl group. Such is not the case because aconitase discriminates between the two arms of citrate.

e. False. Malate can easily be dehydrated to fumarate by reversal of the fumarase reaction.

13. a. Without malate synthase, the yeast would be unable to grow on 2-carbon precursors as sole carbon source because TCA cycle intermediates could not be synthesized. The glyoxalate bypass provides a pathway for the net synthesis of malate from acetyl-CoA. Entry of malate, a dicarboxylate intermediate in the TCA cycle, would provide increased concentrations of each of the TCA cycle intermediates with subsequent increase in oxidation of acetyl groups. ATP formation by oxidative phosphorylation could then occur. In addition, the glyoxalate bypass provides a carbon source of synthesis of carbohydrate and precursors of other cellular constituents. The absence of malate synthase would adversely affect these reactions.

b. Pyruvate carboxylase catalyzes the carboxylation of pyruvate to form oxaloacetate. Oxaloacetate thus formed enters the TCA cycle to replenish cycle intermediates. Pyruvate carboxylase is normally activated by acetyl-CoA; thus the need for oxaloacetate is tied to an increased supply of acetyl-CoA. Were the pyruvate carboxylase in the mutant less responsive to acetyl-CoA, it is possible that the TCA cycle activity would markedly diminish if the cycle intermediates were being used in biosynthetic pathways. In addition, the biosynthetic pathways (lipids, amino acids, and carbohydrates) dependent on TCA cycle intermediates would also be inhibited due to lack of metabolites.

c. If the PDH were inhibited more strongly than usual by acetyl-CoA, one might suspect that acetyl-CoA concentration in the mitochondrial matrix would markedly decrease, in turn limiting activity of citrate synthase and diminishing TCA cycle activity. Hence the organism may become growth-limited because of lowered energy production and because of diminished concentrations of biosynthetic precursors supplied by the TCA cycle.

15. a. Fumarate + NADH + H^+ $\rightarrow$ succinate + NAD^+.

b. Fumarate is formed from the oxidation of succinate, a TCA cycle intermediate formed from the decarboxylation of α-ketoglutarate. α-Ketoglutarate dehydrogenase is an integral enzyme in the TCA cycle and is active aerobic metabolism of acetyl-CoA. However, under anaerobic conditions in *E. coli*, there is little, if any, α-ketoglutarate dehydrogenase activity. Succinate, and therefore fumarate, are not produced by this route. Some reactions of the TCA cycle are reversible, however. For example, the reduction of oxaloacetate, formed by carboxylation of PEP pyruvate, yields L-malate, which may be dehydrated to fumarate. These reversible reactions form fumarate even if succinate dehydrogenase or α-ketoglutarate dehydrogenase reactions are blocked. (See Spiro, S., and J. R. Guest, *TIBS.* 16:310–314 (1991).) Consider the reactions

$$\text{Phosphoenolpyruvate} + CO_2 \rightarrow \text{oxaloacetate} + \text{Pi}$$
(PEP carboxylase)

$$\text{Oxaloacetate} + \text{NADH} + H^+ \rightarrow \text{L-malate} + NAD^+$$
(malate dehydrogenase)

$$\text{L-Malate} \rightarrow \text{fumarate} + \text{HOH}$$
(fumarase)

$$\text{Fumarate} + \text{NADH} + H^+ \rightarrow \text{succinate} + NAD^+$$
(fumarate reductase)

c. In the reactions shown in part (6b), four reducing equivalents (two hydride groups) are transferred to carbon acceptors. Reduction of oxaloacetate to malate by malate dehydrogenase (MDH) and reduction of fumarate to succinate by fumarate reductase each requires hydride (or equivalent) transfer from NADH to the organic substrate. In the reduction of pyruvate to lactate via LDH only one hydride is used. Thus, two equivalents of NAD^+ are resupplied to glycolysis by the activities of MDH and fumarate reductase, whereas only one equivalent of NAD^+ is regenerated by LDH. Fumarate is one of the terminal electron acceptors used during anaerobic respiration in *E. coli.* However, each mole of PEP carboxylated is at the expense of 1 mole equivalent of ATP that could have been formed as a product of pyruvate kinase.

17. The committed step in a metabolic pathway is usually under metabolic control. Inhibition of the committed step in a metabolic sequence or pathway prevents the accumulation of unneeded intermediates and effectively precludes activity of the enzymes using those intermediates as substrates. The decarboxylation of pyruvate and the oxidative transfer of the hydroxyethyl group by pyruvate dehydrogenase constitutes the committed step in the pyruvate dehydrogenase catalytic sequence and is a logical control point.

14 Electron Transport and Oxidative Phosphorylation

Summary

In eukaryotes, most of the reactions of aerobic energy metabolism occur in mitochondria. An inner membrane separates the mitochondrion into two spaces: the internal matrix space and the intermembrane space. An electron-transport system in the inner membrane oxidizes NADH and succinate at the expense of O_2, generating ATP in the process. The operation of the respiratory chain and its coupling to ATP synthesis can be summarized as follows:

1. Electron transfer to O_2 occurs stepwise, through a series of flavoproteins, cytochromes (hemeproteins), iron-sulfur proteins, and a quinone. Most of the electron carriers are collected in four large complexes, which communicate via two mobile carriers—ubiquinone (UQ) and cytochrome *c*. Complex I transfers electrons from NADH to UQ, and complex II transfers electrons from succinate to UQ. Both of these complexes contain flavins and numerous iron-sulfur centers. Complex III, which contains three cytochromes (cytochromes b_L, b_H, and c_1) and one iron-sulfur protein, passes electrons from reduced ubiquinone (UQH_2) to cytochrome *c*. Complex IV contain two cytochromes (a and a_3) and two Cu atoms, and transfers electrons from cytochrome *c* to O_2. The transfer of electrons through complex III occurs by a cyclic series of reactions (the Q cycle), in which UQH_2 and UQ undergo oxidation and reduction at two distinct sites.
2. As electrons move through complexes I, III, and IV, protons are taken up from the matrix and released on the cytosolic side of the membrane. This raises the pH of the matrix and leaves the matrix negatively charged relative to the cytosol, creating an electrochemical potential difference that tends to pull protons from the cytosol back into the matrix.
3. Proton influx through the F_o base-piece of the ATP-synthase in the inner membrane drives the formation of ATP by causing the release of bound ATP from the catalytic site on the F_1 head-piece of the enzyme. Approximately 2.5 molecules of ATP are synthesized for each pair of electrons that pass down the electron-transport chain from NADH to O_2.
4. Uncouplers dissipate the electrochemical potential gradient by carrying protons across the membrane; respiration then occurs rapidly even in the absence of phosphorylation.
5. The electrochemical potential gradient also drives the uptake of P_i and ADP into the mitochondrial matrix and the export of ATP to the cytosol.

Problems

1. Compare and contrast the role of heme in hemoglobin-myoglobin versus the cytochromes.

2. In the reaction catalyzed by dihydrolipoyl dehydrogenase, one of three enzymes in the pyruvate dehydrogenase complex (fig. 13.5), electrons flow from oxidized lipoic acid to enzyme-bound FAD to NAD^+. Compare the flow of electrons in the latter part of this scheme (FAD to NAD^+)

to the flow of electrons in the electron transport scheme (Complex I). Is there a distinct difference in the flow of electrons in the two schemes? If so, can you provide a possible explanation for this difference?

3. Calculate the ATP yield (mole/mole) of the complete oxidation of each of the following: (a) pyruvate, (b) glyceraldehyde-3-phosphate, (c) the acetyl portion of acetyl-CoA.

4. Why have the classical methods of enzymology (purification and characterization of proteins) been of limited help in deducing how the electron-transport system functions?

5. Why did nature pick iron, and to a much lesser degree copper, rather than calcium, magnesium, potassium, and so forth for roles in the electron-transport scheme?

6. Compare iron-sulfur proteins, flavoproteins, and quinones with respect to the following:
   a. Chemical nature of the functional group that undergoes oxidation-reduction.
   b. Number of reducing equivalents per redox center involved in electron donor/acceptor reactions of physiological importance. Indicate if semiquinones are formed and include them in the reduction scheme.
   c. Stoichiometry of protons taken up per electron.

7. a. Describe how heme is bound to the protein portion of the a/a_3-, *b*-, and *c*-type cytochromes.
   b. Although three types of cytochromes occur in rat liver mitochondria, CO and CN^- inhibit electron transfer only at the cytochrome a/a_3 complex. Why do these inhibitors interact with cytochrome a/a_3 but not with cytochrome *b* or cytochrome *c*?

8. Calculate the standard redox potential change ($\Delta E^{\circ\prime}$) and the standard free energy change ($\Delta G^{\circ\prime}$) for the following reactions at pH 7.0. Write a balanced equation for each reaction.
   a. Cyt *c* (Fe^{2+}) + cyt a_3 (Fe^{3+}) $\rightarrow$cyt *c* (Fe^{3+}) + cyt a_3 (Fe^{2+})
   b. 4 cyt *c* (Fe^{2+}) + O_2 + $4H^+$ $\rightarrow$ 4 cyt *c* (Fe^{3+}) + 2 HOH
   c. Oxidation of succinate by succinate: Cytochrome *c* reductase.

9. a. In biological oxidation–reduction reactions, does the stoichiometry of electron transfer (reducing equivalents per mole) differ among the 1Fe, 2Fe-2S, and 4Fe-4S centers?
   b. 4Fe-4S centers function in electron transport over a wide range of reduction potentials. Nothing inherent in the iron–sulfur cluster suggests this range of reduction potentials. Therefore, what other component(s) must dictate reduction potential?

10. What percentage of cytochrome *c* will be in the oxidized form in a solution held at +0.30 V and pH 7.0?

11. Given the standard reduction potentials for cytochrome *c* and ubiquinone at pH 7.0 (see text), calculate the corresponding values at pH 6.0 and 8.0.

12. Rotenone, which blocks the transfer of electrons from $FMNH_2$ of the NADH dehydrogenase to ubiquinone, is a potent insecticide and fish poison.
   a. Explain why rotenone is lethal to insects and fish.
   b. Would you expect the use of rotenone as an insecticide to be potentially hazardous to other animals (e.g., humans)? Why or why not?
   c. If isolated mitochondria are respiring with succinate as substrate, do you expect a change in O_2 consumption on addition of rotenone? If β-hydroxybutyrate is the respiratory substrate?

13. a. Explain the necessity of having a ubiquinone concentration in excess of other mitochondrial electron-transfer components.
b. Suppose that you are examining muscle tissue mitochondria in which the UQ content is well below normal. You find that concentrations of all other electron-transfer components are within the normal range. Predict the effect of UQ deficiency on oxidation of: (i) NADH-producing substrates, (ii) succinate, (iii) ascorbate plus a redox mediator.
c. Explain why UQ-deficient muscle tissue has greater than normal concentrations of lactate. (Ogasahara, Engel, Frens, et al., *Proc. Natl. Acad. Sci. USA* 86:2379–2382, 1989.)

14. a. Explain what is meant by "tightly coupled" mitochondria. How can we determine whether mitochondria are tightly coupled?
b. What is the importance of "respiratory control" in oxidation of metabolites?
c. In what metabolic circumstance is it advantageous for the organism to have mitochondria uncoupled?

15. The uncoupling reagent, 2,4-dinitrophenol (2,4-DNP) is highly toxic to humans, causing marked increase in metabolism, body temperature, profuse sweating, and in many instances, collapse and death. For a brief period in the 1940s, however, doses of 2,4-DNP presumed to be sublethal were prescribed as a means of weight reduction in humans.
a. Explain why administration of 2,4-DNP results in increased metabolic rate as evidenced by increased O_2 consumption.
b. How are the metabolic events resulting from administration of 2,4-DNP pertinent to regulation of glycolysis and the TCA cycle?
c. Why does consumption of 2,4-DNP lead to hyperthermia and profuse sweating?
d. Explain how 2,4-DNP uncouples oxidative phosphorylation.

16. A suspension of mitochondria is incubated with pyruvate, malate, and ^{14}C-labeled triphenylmethyl-phosphonium [TPP] chloride under aerobic conditions. The mitochondria are rapidly collected by centrifugation, and the amount of ^{14}C that they contain is measured. In a separate experiment, the volume of the mitochondrial matrix space was determined so that the concentration of TPP cation in the matrix can be calculated. The internal concentration is found to be 1,000 times greater than that in the external solution.

a. What is the apparent electric potential difference ($\Delta\Psi$) across the inner membrane? Express your answer in the appropriate units, and indicate which side of the membrane is positive.
b. Qualitatively, how might $\Delta\Psi$ be affected by the addition of an uncoupler?

17. Differentiate between electrogenic and neutral transport systems in mitochondria. How is electrogenic transport influenced by the membrane potential? What is the effect of neutral transport on the pH gradient?

Solutions

1. There are several fundamental differences. One is that in hemoglobin/myoglobin the heme serves as a carrier with oxygen-heme iron (Fe^{2+}) interacting in a ligand-metal coordination relationship. In the cytochromes heme serves in a redox role with the iron interconverting between Fe^{2+} and Fe^{3+}. The heme moiety is held in place in hemoglobin/myoglobin by non-covalent forces while in many cytochromes (e.g., cytochrome) the heme moiety is covalently bound to the protein.
3. See also table 14.3 ATP (mole/mole) produced from sources listed on left using the starting materials listed below.

	(a) pyruvate	(b) glycer-aldehyde-3-phosphate	(c) acetyl portion of acetyl-CoA
Cyto NADH	0	1.5	0
Cyto ATP	0	2	0
Mito NADH	10	10	7.5
Mito ATP (GTP)	1	1	1
Mito $FADH_2$	1.5	1.5	1.5
Sum	12.5	16	10

5. Both iron and copper have two common stable ions; the transition between Fe^{2+}/Fe^{3+} or Cu^{1+}/Cu^{2+} are both redox reactions. By placing these ions (mainly iron) in different environments a wide variety of $E^{o\prime}$ have been achieved by nature. The other metals listed do not have two stable ions.
7. a. Heme of the mitochondrial *b*-type cytochrome interacts hydrophobically with adjacent hydrophobic residues from the membrane-spanning α helices. The heme iron is fully coordinated through two imidazole groups from histidines in the protein. Heme in the *c*-type

cytochromes is covalently bound to the protein through thioethers formed by the addition of cysteinyl sulfhydryl groups to the vinyl substituents on the heme ring. The iron is also fully liganded to a nitrogen from the imidazole group of histidine and a sulfur from the thioether linkage of methionine providing the fifth and sixth ligands. Heme *a* differs from the protoheme (heme) of the *b*- and *c*-type cytochromes by substitution of a formyl group at ring position 8 and a 17-carbon isoprenoid chain at position 2. The hydrophobic isoprenoid chain provides added hydrophobic interaction between the heme *a* and the protein. Iron of heme *a* is fully coordinated, but heme a_3 has an open ligand position available for binding of oxygen.

b. Carbon monoxide binds to the reduced heme a/a_3 and cyanide binds to the oxidized heme a/a_3 presumably at the oxygen-binding site; they inhibit transfer of electrons from reduced cytochrome *c* to O_2. Electron transport is effectively blocked and ADP phosphorylation ceases. Neither CO nor CN^- at low concentration interact with the cytochrome *b* or *c* heme iron because there is no open ligand position to the iron available in these heme proteins.

9. a. In biological systems, the iron-sulfur centers are obligatorily single electron donors/acceptors, regardless of the number of Fe atoms in the center or their initial oxidation state. For example, the 4 (Fe-S) cluster iron-sulfur proteins accept electrons from flavoproteins or quinones in only 1-electron transfer cycles.

b. The 4Fe-4S cluster is found in iron-sulfur proteins that transfer electrons at low and at high potential. The reduction potential is a measure of the ease of addition of an electron to the couple, compared to the standard hydrogen electrode. Thus the protein component of the iron-sulfur protein affects the reduction potential. For example, an electron-withdrawing environment at the Fe-S cluster in principle should yield a more positive reduction potential than if the Fe-S cluster were in an electron-donating environment. Spatial constraints and physical interaction between the protein and the iron-sulfur cluster may also affect the redox potential. Moreover, the initial redox state of the cluster affects reduction potential, for example, the oxidized form of high potential iron-sulfur protein cluster is (3 Fe^{3+} –1 Fe^{3+}) whereas the oxidized form of the low potential iron-sulfur cluster is (2 Fe^{2+} –2 Fe^{3+}).

11. The standard reduction potential $E^{\circ\prime}$ changes by the ratio $\Delta E'/\Delta pH = -(60\text{ mV})(n^{H}+/n)$ where $n^{H}+$ is the proton equivalents taken up per equivalent (n) of electrons transferred in the reaction.

$$E^{\circ\prime} = E^{\circ\prime}\text{ (pH 7)} - E^{\circ}\text{ (pH)}$$

The standard reduction potential of cytochrome *c* is unaltered at either pH 6 or pH 8 because no protons are taken up in the reduction reaction. The iron in the heme is reduced from ferric to ferrous state. Reduction of ubiquinone results in the addition of two equivalents of protons per 2 electron equivalents in reducing the quinone to the dihydroquinone. Thus at pH 6, ΔpH is (7 – 6) = +1, so

$$E^{\circ\prime} - E^{\circ}\text{ (pH 6)} = -60\text{ mV}/1$$
$$E^{\circ\prime}\text{ (pH 7)} - E^{\circ}\text{ (pH 6)} = -60\text{ mV}$$

$$E^{\circ}\text{ (pH 6)} = E^{\circ\prime} + 60\text{ mV} = (110\text{ mV} + 60\text{ mV}) = +170\text{ mV}$$
$$E^{\circ}\text{ (pH 6) is } + 170\text{ mV}$$

At pH 8,

$$\begin{aligned} [E^{\circ\prime} - E^{\circ}\text{ (pH 8)}] &= -60\text{ mV}/-1 \\ E^{\circ\prime} - E^{\circ}\text{ (pH 8)} &= +60\text{ mV} \\ E^{\circ}\text{ (pH 8)} &= E^{\circ\prime} - 60\text{ mV} \\ &= (110\text{ mV} - 60\text{ mV}) \\ E^{\circ}\text{ (pH 8)} &= 50\text{ mV} \end{aligned}$$

Consider the reaction $UQ + 2e^- + 2H^+ \rightarrow UQH_2$. Decreasing pH (increasing H^+ concentration) favors the formation of UQH_2 by mass action. Hence the quinone should be more easily reduced to UQH_2 at pH 6 and the value of $E°$ calculated is consistent with the prediction. Decreasing the proton concentration favors the oxidized form of the quinone again by mass action and the quinone should be more difficult to reduce.

13. a. Ubiquinone (coenzyme Q) is the electron acceptor for a number of dehydrogenases in the mitochondrial electron-transport system, including succinate dehydrogenase, NADH dehydrogenase, the flavoprotein α-glycerol phosphate dehydrogenase and the electron transfer flavoprotein dehydrogenase. Ubiquinol is oxidized by complex III. The large amount of UQ is necessary to ensure efficient transfer of electrons from the mitochondrial dehydrogenases to complex III. Although the ratio of UQ to Complex III is approximately 20:1, diffusion of UQ among the various dehydrogenases may be the rate-limiting step in electron transfer. (See Hackenbrock, C. R. *TIBS*, 6:151–54, 1981.)

    b. (i) If UQ (ubiquinone) were limiting, the rate of oxidation of NADH by the mitochondrial electron-transfer system would also be limited. NADH concentration would increase, the NAD^+ supply would decrease, and the NAD^+-dependent dehydrogenases would be inhibited. Subsequently, the rate of oxidation of NADH-producing substrates would decrease.

    (ii) Electrons from succinate oxidation are also transferred to ubiquinone from the succinate dehydrogenase, so a deficiency of the quinone would limit succinate oxidation.

    (iii) Ascorbate plus a redox mediator reduces cytochrome *c* but does not transfer electrons to ubiquinone. Limiting amounts of the quinone should not affect ascorbate-dependent reduction of cytochrome *c*.

    c. Lactate is formed in skeletal muscle by the NADH-dependent reduction of pyruvate. In resting muscle, the NADH formed by the glycolytic enzyme glyceraldehyde-3-phosphate dehydrogenase should be oxidized in the mitochondria via one of the reducing equivalent shuttles (α-glycerophosphate or malate). The deficiency of UQ will decrease the rate of extramitochondrial NADH oxidation resulting in an increase in the amount of pyruvate reduced to lactate. The lactate content would be expected to increase rapidly upon mild exercise.

15. a. The uncoupler 2,4-dinitrophenol circumvents respiratory control in the mitochondria by short-circuiting the proton gradient. The lipophilic weak acid transports H^+ across the membrane, bypassing the F_1-F_0 complex. Substrates will be oxidized independently of ADP or ATP concentrations, and O_2 reduction will be more rapid than in state 3 respiration. Although respiration is rapid, no ADP is phosphorylated.

    b. Uncoupled mitochondria oxidize NADH and succinate but fail to phosphorylate ADP. Cellular processes will continue to utilize ATP, causing an accumulation of ADP and AMP. Increased levels of NAD^+ and ADP or AMP activate both the TCA cycle and glycolysis. The rate of oxidation of carbohydrates and fatty acids would markedly increase.

    c. Oxidation of substrates with the subsequent reduction of oxygen releases 52 kcal per mole of NADH oxidized (or about 39 kcal per mole of succinate oxidized). The energy normally would be used in part to drive the vectorial accumulation of H^+ and subsequent phosphorylation of ADP. The mitochondria are uncoupled and use little, if any, energy to phosphorylate ADP. The energy is dissipated as heat, leading to elevated both temperature (hyperthermia) and profuse perspiration in an effort to decrease body temperature. The metabolic scenario is reminiscent of the brown fat mitochondria and nonshivering thermogenesis discussed in solution 11c.

d. 2,4-Dinitrophenol is a lipid soluble weakly acidic compound thought to allow equilibration of protons across the inner mitochondrial membrane. Although protons are translocated from the matrix across the inner membrane during electron transfer, the proton gradient would immediately be depleted without passing through the F_O- F_1-dependent ADP phosphorylation system. Respiratory (ADP) control would be lost.

17. The inner mitochondrial membrane contains specific transmembrane proteins that transport charged molecules across the inner mitochondrial membrane. Charged molecules diffuse with little restriction across the outer membrane. The transport process is electrogenic if the export of one molecule coupled with the import of another molecule yields a net charge difference across the membrane. In general terms, transfer of A^{3-} from the matrix and A^{3-} into the matrix yields a net negative charge on the cytoplasmic side of the membrane. Electrogenic processes are driven by the membrane potential ($\Delta\Psi$). Transport that diminishes the membrane potential are energetically favorable and are driven in that direction. H^+-transport driven by the electron-transport system is another example of electrogenic (active) transport. Proton translocation increases the net charge differential across the membrane and energy is required to drive the transport.

    Neutral transport processes exchange molecules of net identical charge (sign and magnitude) in the opposite vectorial direction or oppositely charged molecules in the same vectorial direction. For example, exchange of a dianions (succinate for malate) is a neutral transport process. Neutral processes in the mitochondria may be driven by the H^+ gradient, for example, cotransport of H^+ and A^-. Processes coupled to export of OH^- are equivalent to H^+ import, that is, to an energetically favorable decrease in the proton gradient.

15 Photosynthesis

Summary

1. Photosynthesis, the conversion of sunlight into chemical energy, occurs in algae and some bacteria, in addition to higher plants. In plants, the photochemical reactions of photosynthesis take place in the chloroplast thylakoid membrane. In bacteria, they occur in the plasma membrane.
2. Photosynthetic organisms take advantage of the fact that chlorophyll becomes a strong reductant when it is excited with light. Photooxidation of chlorophyll or bacteriochlorophyll occurs in pigment-protein reaction centers. Chloroplasts have two types of reaction centers: Photosystem I, which contains the reactive chlorophyll complex P700, and photosystem II, which has P680. The reactive complex in purple photosynthetic bacteria (P870) is a bacteriochlorophyll dimer.
3. Photosynthetic membranes also contain pigment-protein complexes that serve as antennas. When the antenna absorbs a photon, energy hops rapidly from complex to complex by resonance energy transfer until it is trapped in a reaction center.
4. When P870 is excited, it transfers an electron to a bacteriopheophytin. The bacteriopheophytin then reduces a quinone, which in turn reduces another quinone, Q_B. A cytochrome replaces the electron released by P870. When Q_B receives a second electron, it picks up protons from the cytosol. Electrons return from Q_BH_2 to the cytochrome via a cytochrome bc_1 complex, and protons are released outside the cell. This pumping of protons creates an electrochemical potential gradient across the plasma membrane. The return of protons to the cell through an ATP-synthase drives synthesis of ATP.
5. Photosystems I and II of chloroplasts operate in series. Photooxidation of P680 in photosystem II generates a strong oxidant that can oxidize H_2O to O_2. In this process, a manganese complex progresses through five different oxidation states (S_0–S_4). When the complex reaches state S_4, O_2 is given off. The electron released by P680 goes to a pheophytin and then to two plastoquinones. When the second plastoquinone has been doubly reduced, it picks up protons from the chloroplast stromal space. Electrons move from the reduced plastoquinone to P700 of photosystem I via the cytochrome b_6f complex and a copper protein (plastocyanin).
6. Photooxidation of P700 in photosystem I reduces a chlorophyll, which transfers electrons to a series of membrane-bound iron–sulfur centers, probably by way of a quinone. From the iron–sulfur centers, electrons move to the soluble iron–sulfur protein, ferredoxin, and then to a flavoprotein that reduces $NADP^+$. Cyclic electron flow may also occur from the iron–sulfur proteins to carriers between the two photosystems.
7. Electron flow through the cytochrome b_6f complex results in proton translocation from the stroma to the thylakoid lumen. In addition, protons are released in the lumen when H_2O is oxidized and are taken up from the stromal space when $NADP^+$ is reduced. Protons move from the thylakoid lumen back to the stroma through an ATP-synthase, driving the formation of ATP.
8. ATP and NADPH are used for incorporation of CO_2 into carbohydrates. CO_2 reacts with ribulose-1,5-bisphosphate to give two molecules of 3-phosphoglycerate, under the influence of ribulose bisphosphate carboxylase. 3-Phosphoglycerate can be converted back to ribulose-1,5-bisphosphate at the expense of ATP and NADPH in the reductive pentose cycle. For every 3 molecules of CO_2 that enter the cycle, there is a gain of one glyceraldehyde-3-phosphate.

Ribulose bisphosphate carboxylase also catalyzes a wasteful oxygenation reaction of ribulose-1,5-bisphosphate. Some species of plants use an auxiliary cycle of reactions (the C-4 cycle) to concentrate CO_2 in the cells that carry out the reductive pentose cycle, thus allowing the plants to fix CO_2 at elevated rates.

Problems

1. The word "fixed" is often used to describe what happens to CO_2 in photosynthesis or to N_2 in nitrogen "fixation." How do you explain the meaning of "to fix" a gas or "a fixed gas" to an English major?

2. If the conversion of light energy into chemical energy occurred with 100% efficiency, how many moles of ATP could be produced per mole of photons of blue light (450 nm) and red light (700 nm)?

3. What arguments can you make for or against the idea that the chlorophylls (fig. 15.4) are aromatic compounds?

4. The lowest energy absorption band of P870 occurs at 870 nm.
   a. Calculate the energy of an einstein of light at this wavelength.
   b. Estimate the effective standard redox potential ($E^{\circ\prime}$) of P870 in its first excited singlet state, given that the $E^{\circ\prime}$ for oxidation in the ground state is +0.45 V.

5. The traces shown here are measurements of optical absorbance changes at 870 and 550 nm when a suspension of membrane vesicles from photosynthetic bacteria was excited with a short flash of light. Downward deflection of the traces represent absorbance decreases. Explain the observations. (Absorption spectra of a *c*-type cytochrome in its reduced and oxidized forms are described in the previous chapter.)

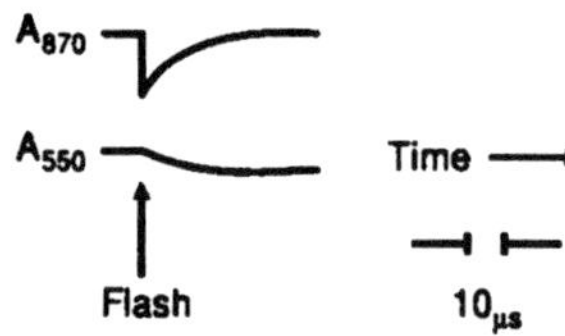

6. You add a nonphysiological electron donor to a suspension of chloroplasts. When you illuminate the chloroplasts, the donor becomes oxidized. How can you determine whether this process involves both photosystems I and II? (In principle, the donor can transfer electrons either to some component on the O_2 side of photosystem II or to a component between the two photosystems.)

7. Ubiquinone has an absorbance band at 275 nm. This band bleaches when the quinone is reduced to either the semiquinone or the dihydroquinone. The anionic semiquinone has an absorption band at 450 nm but neither the quinone nor the dihydroquinone absorbs at this wavelength. A suspension of purified bacterial reaction centers was supplemented with extra ubiquinone and reduced cytochrome *c* and was then illuminated with a series of short flashes of light. The absorbance at 275 nm decreased on the odd-numbered flashes as shown on the first curve below. The absorbance at 450 nm increased on the odd-numbered flashes but returned to the original level on the even-numbered flashes as shown in the second curve.

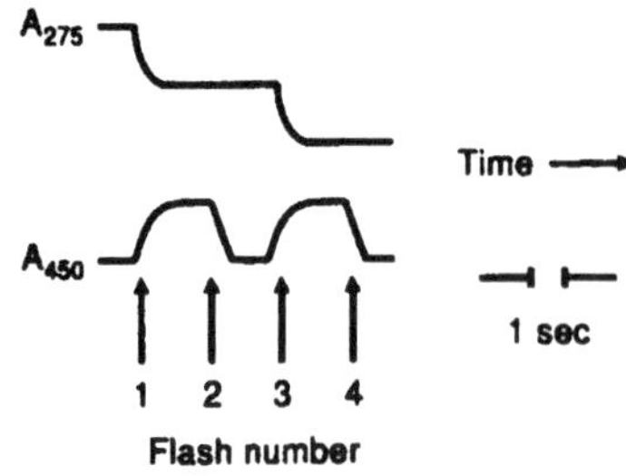

a. Explain the patterns of absorbance changes at the two wavelengths.
b. Why is it necessary to have reduced cytochrome *c* present in order to see these effects?

8. Explain why in green plants, the quantum efficiency of photosynthesis drops sharply at wavelengths longer than 680 nm.

9. Do you predict that plant cells lacking carotenoids are more or less susceptible to damage by photooxidation? Explain.

10. Molecular oxygen is an alternative substrate for the ribulose bisphosphate carboxylase–oxygenase and is also a competitive inhibitor with respect to CO_2 fixation. Explain.

11. In the formation of phosphoenolpyruvate from pyruvate in the mesophyll cells (fig. 15.28), notice the energetics of the reaction. How does this process compare energetically with the formation of phosphoenolpyruvate from pyruvate in the start of gluconeogenesis?

12. Provide a mechanism for the conversion of pyruvate to phosphoenolpyruvate in the mesophyll cells (fig. 15.28), and account for the consumption of an ATP and P_i and the production of AMP, PP_i, and phosphoenolpyruvate.

13. Propose a mechanism for the "oxygenase" activity of ribulose bisphosphate carboxylase.

14. When examining the information described in figures 15.13 or 15.17, one is struck by the similarities with the electron-transport scheme (chapter 14). What types of components do these two schemes have in common?

15. What differences can you find between the pentose phosphate pathway (figs. 12.31 through 12.34 and associated text) and the Calvin cycle?

16. Reduction of 3 moles of CO_2 to form 1 mole of triose phosphate requires 9 moles of ATP and 6 moles of NADPH.
    a. What is the source of NADPH in the reduction of CO_2?
    b. Account for the ATP consumed in the formation of triose phosphate.
    c. Assume that the CO_2 initially is added to PEP in the C-4 pathway. What is the additional cost in ATP per mole of CO_2 added?
    d. Is additional NADPH required for CO_2 fixation in the C-4 plant?

Solutions

1. The word fixed refers to converting a gas into a liquid or solid. Much of the early chemistry (18th century) was done with gases because they could be produced in a pure form by various reactions. The only reliable method for quantifying materials in this time period was by weighing. To weigh a gas it was converted into a solid or liquid by a chemical reaction and the resulting solid or liquid was weighed. Because the gas was "no longer moving about" it was "fixed."
3. Chlorophyll molecules are flat, completely conjugated ring compounds that obviously have a resonance form. Chlorophylls have a ring containing 9 pi bonds or 18 pi electrons which fits the n = 4 situation in the Huckel (4n + 2) rule. The (cis, trans, trans)$_3$ double bond pattern of the chlorophylls also fits that of [18]annulene which is known to be aromatic.

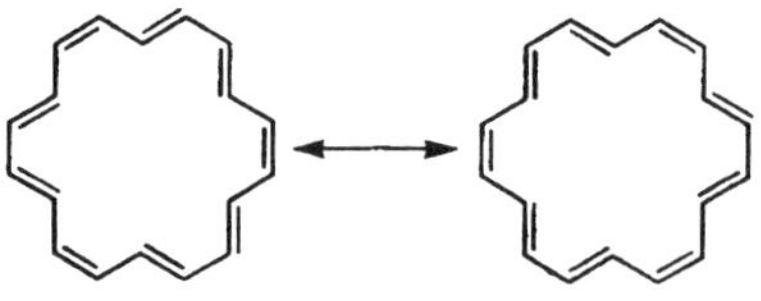

[18]annulene

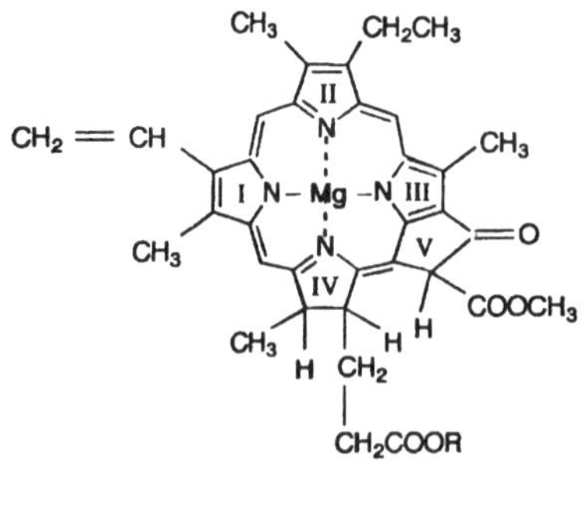

CH3 CH2CH3 CH2 = CH CH3 I N- Mg -N III II IV V =O CH3 COOCH3 CH3 H CH2 H H CH2COOR

5. Upon illumination, the chromatophore P_{870} is activated by absorption of a photon of light. The absorbance at 870 nm decreases because the pi-cation radical of the oxidized chromatophore has a lower absorbance at that wavelength. Thus, the trace monitoring 870 nm decreases upon illumination of the chromatophore. The *c*-type cytochrome is added initially in the reduced form (cyt c^{2+}) and absorbs at 550 nm. Oxidized cytochrome *c* (cyt c^{3+}) has only a small absorbance at 550 nm. Electron transfer from reduced cytochrome *c* to the pi-cation radical (P_{870+}) regenerates the ground state P_{870} and forms oxidized cytochrome *c*. The absorbance of the P_{870} increases to the initial level and the absorbance at 550 nm of the cytochrome *c* pool decreases. The explanation is consistent with the observed upward trace at 870 nm and the downward trace at 550 nm over 30 μs.
7. a. Illumination of the bacterial photocenter generates an active reductant that transfers an electron to ubiquinone, forming the semiquinone. Some, but not all, of the ubiquinone is reduced by a single flash. The absorbance at 275 nm decreases because the concentration of ubiquinone is diminished and because the semiquinone radical does not absorb 275 nm light. The absorbance increase at 450 nm is consistent with the formation of semiquinone radical that absorbs 450 nm light. The second flash activates transfer of a second electron from the photocenter to reduce the bound semiquinone to dihydroquinone. The dihydroquinone absorbs at neither 275 nm nor 450 nm. Reduction of the semiquinone form abolishes the absorbance at 450 nm. The decrease in absorbance at 275 nm is consistent with the decrease in oxidized (ubiquinone) concentration. The sequence of events is repeated with the third and fourth flashes.
   b. Transfer of an electron from the activated photocenter (P_{870}*) to the ubiquinone leaves the oxidized pi-cation radical form of the reaction center (P_{870}+), which must be reduced to allow a second light-activation cycle. In this experiment, reduced cytochrome *c* is the electron donor to P_{870}+. Were the reductant omitted, the P_{870}+ might oxidize the reduced quinone, or the activated reaction center P_{870}* might return to ground state by emission of fluorescence.
9. Oxygen, a product of photosynthesis, may damage the chloroplast if the ground-state (triplet) oxygen is activated to the singlet state. Single oxygen may be formed by activated (singlet) chlorophyll that decays to the triplet state. Whereas energy transfer between singlet-state chlorophyll and triplet-state oxygen does not occur, triplet-state chlorophyll may activate triplet-state oxygen to the reactive singlet state. Singlet-state oxygen causes cumulative oxidative damage to the chloroplast. Carotenoids compete with the oxygen for the triplet-state chlorophyll and inhibit singlet oxygen production. Plants deficient in carotenoids risk photooxidative damage because of an increased flux of singlet oxygen. Although there are enzymes that scavenge the reduced, activated oxygen intermediates (superoxide, hydrogen peroxide), there is no enzymatic activity thus far identified that catalytically scavenges singlet oxygen.
11. The mechanisms are different but the energetic outcome is the same. In the start of gluconeogenesis pyruvate is converted to phosphoenolpyruvate with the consumption of two equivalents of "ATP" (ADP = product). In the phosphoenolpyruvate production in mesophyll cells the conversion involves the production of an AMP from an ATP. Considering the ubiquitous pyrophosphatase, this is the energetic equivalent of the conversion of 2 ATP to 2 ADP.

13. The enolate intermediate (fig. 15.26) reacts with oxygen to produce a cycloperoxide intermediate which decomposes into glycerate-3-phosphate and glycolate-2-phosphate (fig. 15.27).

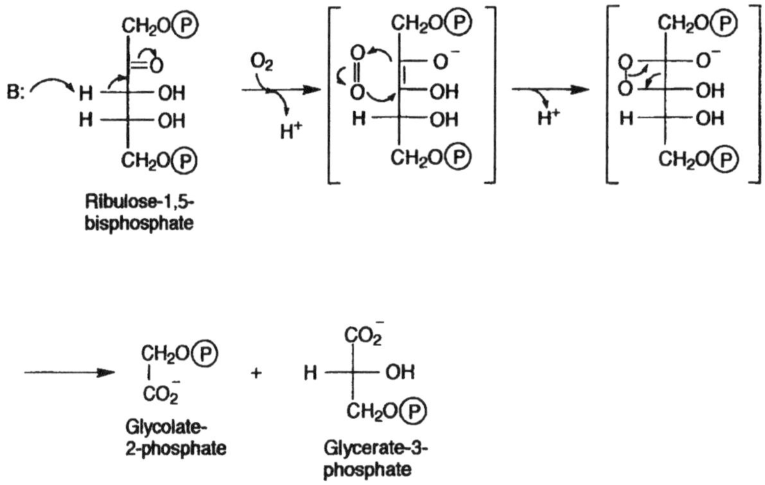

Ribulose-1,5-bisphosphate

Glycolate-2-phosphate

Glycerate-3-phosphate

15. The Calvin Cycle uses many of the metabolites and enzymes found in the pentose phosphate pathway and glycolysis. Some of the many differences noted include: The involvement of erythrose-4-phosphate and dihydroxyacetone phosphate in the production of sedoheptulose-1,7-bisphosphate, phosphorylation of ribulose-5-phosphate to ribulose-1,5-bisphosphate, the addition of CO_2 to ribulose-1,5-bisphosphate and the production of two glyceraldehyde-3-phosphates. Also notice that the coenzyme specificity for the glyceraldehyde-3-phosphate dehydrogenase is $NADP^+$ while the corresponding enzyme in glycolysis is specific for NAD^+.

16 Structures and Metabolism of Oligosaccharides and Polysaccharides

Summary

This chapter focuses on the synthesis of complex carbohydrates. It begins with a consideration of the hexoses that are the building blocks of complex carbohydrates. Then some aspects of the synthesis of simple homopolysaccharides are examined followed by a brief consideration of heteropolymers that contain more than one hexose. A major portion of the chapter is concerned with glycoproteins that contain complex linear and branched carbohydrates attached to proteins. Finally the synthesis of the bacterial wall is examined.

1. Hexoses, which are the primary building blocks of oligosaccharides and polysaccharides, come in a large variety of types. All hexoses can be derived from glucose through a series of conversions. These conversions usually occur at the level of the monophosphorylated sugar or the nucleoside diphosphate sugar. The nucleotide sugar is also the activated substrate for formation of disaccharides, oligosaccharides, and polysaccharides.
2. In higher animals a great variety of branched-chain oligosaccharides are found as conjugates in glycolipids and glycoproteins. A great number of sugars and specific glycosyltransferases are involved in oligosaccharide synthesis. Two types of oligosaccharides are distinguished according to their mode of attachment to the protein in the glycoprotein. The O-linked oligosaccharides are synthesized directly on the amino acid side chain hydroxyl group of a serine or a threonine. The N-linked oligosaccharides are synthesized first on a long-chain dolichol phosphate and then transferred to the asparagine side chain of a receptor protein. The protein-attached oligosaccharide is processed by removal of certain sugars and addition of others.

   The synthesis of glycoproteins mostly takes place in two cytoplasmic organelles: the endoplasmic reticulum and the Golgi apparatus. The mature glycoproteins leave the Golgi apparatus in the form of micro-vesicles by budding. For many lysosomal enzymes the oligosaccharide portion is instrumental in targeting the glycoprotein to lysosomes. The oligosaccharide can serve other important recognition functions in addition to ensuring the proper folding of the glycoprotein.
3. The bacterial cell wall contains a heteropolymeric polysaccharide chain that is cross-linked by peptide linkages. The complexity of the resulting peptidoglycan results in part from the complex repeating units and in part from the fact that a cross-linked polymer is made outside the cell. The partially completed polysaccharide structures are transferred from the cytoplasm to extracellular space by attachment to a long-chain undecaprenol lipid that can traverse the cell membrane. This lipid is similar in structure and function to the dolichol phosphate used in oligosaccharide synthesis in animals. Various compounds are effective antibiotics by blocking specific steps in cell wall synthesis. They are very helpful in elucidating the biochemical pathway.

Problems

1. The sequence of amino acids in a protein is determined by the base sequence in DNA. What determines the sequence of sugar residues in oligosaccharides and polysaccharides?

2. The interconversion of glucose and galactose occurs at the UDP-hexose level (fig. 16.2) with a 4-epimerase. The epimerase is unusual in that it has a covalently bound NAD^+. What function do you propose for the NAD^+, and what intermediate do you expect in the reaction?

3. Many amino sugars are found in nature, but glucosamine and galactosamine (figs. 16.1 and 16.2) are by far the most common. By examining their biosynthesis, can you explain why the most common amino sugars in nature are 2-amino sugars?

4. How is lactose made in mothers' milk? What is unusual about the subunit structure of lactose synthase?

5. A great variety of different oligosaccharides result from a limited number of sugars. Explain.

6. How can two different genes for different glycosyl-transferases determine ABO blood group types in humans? Explain why blood group O is considered a universal donor. If you have AB type blood, why can you accept any blood type? A small number of people lack the H antigen (the glycosyltransferase that adds Fucα 1) and have Bombay type blood. What blood type can you give to a person with Bombay type blood and why?

7. Explain why fibroblasts from a patient with I-cell disease secrete lysosomal enzymes when grown in tissue culture.

8. Throughout this chapter notice the occurrence of *N*-acetylglucosamine and *N*-acetylgalactosamine in oligo- or polysaccharides. Why are these sugars present in the *N*-acetyl form?

9. Gluconic acid, as its phosphate derivative, is a metabolite in the pentose phosphate pathway. Why isn't gluconic acid, like glucuronic acid, listed in figure 16.1 as a common component of oligo- or polysaccharides?

10. The hydrophobic amino acid sequence that serves as the binding site for the SRP (fig. 16.8) is on the N terminus of the protein. Why isn't it on the C terminus?

11. Figure 16.17 shows a series of reactions in a cycle. Is this a series of reactions like the tricarboxylic acid cycle or does something actually "go" around the cycle?

12. Compare the pentapeptide portion of the peptidoglycans (fig. 16.17) with a normal sequence of amino acids in a protein. What differences can you identify?

13. What complications have to be overcome to synthesize complex carbohydrates (such as cell wall components and O antigens) outside the cell?

14. Bacteria started for an essential nutrient are not affected by penicillin. Can you explain this?

Solutions

1. The sequence, bond geometry, and linkage type of monosaccharides in an oligosaccharide or polysaccharide is determined by the specificity of the glycosyltransferases involved (see e.g., fig. 16.7 and 16.14).
3. In general amino groups are added to biological molecules in locations once occupied by keto-groups. Because ketoses have carbon-2 as the carbonyl group the majority of the amino sugars in nature are 2-aminosugars. Also, by inspection of figure 16.2 it can be seen that fructose-6-phosphate is the precursor of the amino sugars included in the figure.
5. The large variety of different oligosaccharides results from a limited number of sugars. Sugars have a large number of hydroxyl groups that form glycosidic bonds with the anomeric carbons of other sugars; these anomeric hydroxyl groups can be either α or β conformation. Over eighty different glycosidic linkages are known. With different sugars, number of residues, branching, and other combinations, the number of different oligosaccharides possible is quite large.
7. When fibroblasts from patients with I-cell disease are grown in culture, many lysosomal enzymes are excreted outside the cells. Normally, a lysosomal enzyme is glycosylated and mannose is phosphorylated. There is a mannose-6-phosphate receptor that binds the phosphorylated lysosomal protein and targets it to a transport vesicle that carries the lysosomal enzymes to the lysosome. The genetic defect in I-cell patients is a lack of the enzyme required to phosphorylate mannose. The enzyme thus becomes targeted for secretion outside the cell instead of targeting to the lysosome. If normal lysosomal enzymes (mannose phosphorylated) are placed in the media, these enzymes are taken up by the fibroblast and are targeted to the lysosome, restoring normal lysosome function.
9. If carbon one of glucose is oxidized to a carboxylic acid, gluconic acid results. Gluconic acid cannot form a cyclic hemiacetal and cannot form glucosidic bonds which are required for participation in oligo- or polysaccharides.
11. Figure 16.17 is a series of reactions that can be written in a cyclic format with the undecaprenol phosphate (P-Lipid) functioning as a carrier which is acted upon by the series of reactions. The P-Lipid, therefore, does "go" around the cycle.
13. The major problem in the synthesis of complex carbohydrates outside the cell is the lack of an external energy source such as ATP outside the cell. The biosynthesis of a bacterial cell wall such as the peptidoglycan occurs mainly inside the cell, the final step being the cross-linking of the peptidoglycan occurs mainly inside the cell, the final step being the cross-linking of the

peptidoglycan strands outside the cell. This reaction is a transpeptidation, which does not require any energy source. A peptide bond is broken and another one is formed with the release of alanine. The final step of the synthesis of O-antigens in *Salmonella typhimurium* also occurs outside the cell. In this case the activated forms of the precursors are synthesized inside the cell on a lipid carrier (the same one used in peptidoglycan) and transferred to the outside surface of the cell (held in place by the lipid carrier). The activated galactose of a lipid-linked polymer attacks the mannose of the tetrasaccharide, which is also attached to undecaprenol phosphate (the lipid carrier). In both of these complex reactions to produce extracellular complex carbohydrates a lipid carrier is used to keep intermediates on the surface of the cell. The polymerization is either a transpeptidation (equal bonds broken and formed) or an activated sugar using the lipid carrier.

17 Structure and Function of Biological Membranes

Summary

In this chapter we discussed the structure and function of biological membranes. First we considered their structure starting with an examination of the constituents of membranes. Then we turned to questions concerning the transport of materials across membranes.

1. Biological membranes consist primarily of proteins and lipids whose relative amounts vary considerably.
2. There are two main types of lipids in biological membranes: Phospholipids and sterols. The predominant phospholipids in most membranes are phosphoglycerides, which are phosphate esters of the three-carbon alcohol glycerol.
3. Due to their amphipathic nature, phospholipids spontaneously form ordered structures in water. When phospholipids are agitated in the presence of excess water, they tend to aggregate spontaneously to form bilayers, which strongly resemble the types of structures they form in biological membranes.
4. Membranes contain proteins that merely bind to their surface (peripheral proteins) and those that are embedded in the lipid matrix (integral proteins). Integral membrane proteins contain transmembrane α-helices.
5. Proteins and lipids have considerable lateral mobility within membranes.
6. Biological membranes are asymmetric. Consistent with this asymmetry, a protein that has been inserted into a membrane in a particular orientation usually retains that orientation indefinitely.
7. On heating, phospholipid bilayers undergo a phase transition (melting) from an ordered to a disordered state. The melting temperature (T_m) depends strongly on the phospholipid composition in the bilayer. Increasing the length of the fatty-acid chains increases the T_m; *cis* double bonds decrease the T_m. Cholesterol broadens the melting transition of the phospholipid bilayer. Cells regulate the lipid compositions of their plasma membrane so that a reasonable membrane fluidity is maintained.
8. Some proteins of eukaryotic plasma membranes are connected to the cytoskeleton; this connection inhibits their lateral mobility with the membrane.
9. Biological membranes contain proteins that act as specific transporters of small molecules into and out of the cell. Most solutes are transported by specific carriers that are invariably proteins.
10. Some transporters facilitate diffusion of a solute from a region of relatively high concentration or down a favorable electrochemical potential gradient. Such transporters do not require energy since the transport is in the thermodynamically favorable direction.
11. Other transporters move solutes against an electrochemical potential gradient and require an energy producing process to make them functional. Cells drive such active transport processes in a variety of ways. The transport can be coupled to the hydrolysis of a high energy phosphate, to the cotransport of another molecule down an electrochemical potential gradient, or to the modification of the transported molecule soon after it crosses the membrane.
12. Proteins involved in active transport frequently change their structure during the transport process. For example the NA^+–K^+ pump includes two phosphorylated forms of the enzyme involved in the transport of these two ions.

13. Some membranes contain relatively large pores which allow for the free passage of molecules with molecular weights up to about 600. For example the outer membranes of gram negative bacteria contain pores with diameters of about 10 Å which are formed from proteins called porins.
14. Pores called gap junctions occur in the plasma membranes of cells in some eukaryotic tissues. These channels allow molecules with molecular weights up to about 1000 to diffuse between adjacent cells of the same types.

Problems

1. Examine the list of ingredients on containers of salad dressing. Without much searching you should be able to find lecithin listed. Remember that ingredients are ordered by weight of the item. (The materials that contribute the most are listed first.) After identifying the major components of salad dressing, propose a function for the lecithin.

2. Both triacylglycerol and phospholipids have fatty acid ester components, but only one can be considered amphipathic. Indicate which is amphipathic, and explain why.

3. Phosphatidylcholine usually has a saturated fatty acyl chain on the C-1 of glycerol and an unsaturated fatty acyl chain on the C-2 position. Can you explain how this happens?

4. Why does the Davson-Danielli membrane model predict that the exposed inside of the lipid bilayer is featureless?

5. The relative orientation of polar and nonpolar amino acid side chains in integral membrane proteins is "inside-out" relative to that of the amino acid side chains of water-soluble globular proteins. Explain.

6. What physical properties are conferred on biological membranes by phospholipids? How can the charge characteristics of the phospholipids affect binding of peripheral proteins to the membrane? What role might divalent metal ions play in the interaction of peripheral membrane proteins which phospholipids?

7. Differentiate between peripheral and integral membrane proteins with respect to location, orientation, and interactions that bind the protein to the membrane. What are some strategies used to differentiate between peripheral and integral proteins by means of detergents or chelating agents?

8. Frequently, integral membrane proteins are glycosylated with complex carbohydrate arrays. Explain how glycosylation further enhances the asymmetrical orientation of integral proteins.

9. Integral transmembrane proteins often contain helical segments of the appropriate length to span the membrane. These helices are composed of hydrophobic amino acid residues. In transmembranous proteins with multiple segments that span the membrane, you may find some hydrophilic residue side chains. Why are hydrophilic side chains not favored in single-span membrane proteins? How may the hydrophilic side chains be accommodated in multiple-span proteins?

10. Besides the comments made in the text (page 402) on the free energy changes associated with phosphorylation during sugar transport, what more can be said on the unidirectional aspect of this type of transport system?

11. Elementary portrayals of transport systems have often utilized "revolving doors" as analogies. After reading this chapter what objection(s) do you have with the revolving door idea?

12. Membrane vesicles of *E. coli* that possess the lactose permease are preloaded with KCl and are suspended in an equal concentration of NaCl. It is observed that these vesicles actively, although transiently, accumulate lactose if valinomycin is added to the vesicle suspension. No such active uptake is observed if KCl replaces NaCl in the suspending medium. Explain these results in light of what you know about the mechanism of lactose transport and the properties of valinomycin.

13. The Nernst equation relates the electric potential $\Delta\Psi$ resulting from an unequal distribution of a charged solute across a membrane permeable to that solute, to the ratio between the concentration of solute on one side and on the other:

$$m\Delta\Psi - \frac{-2.3RT}{F}\log\frac{[\mathrm{So}]_1}{[\mathrm{So}]_2}$$

where m is the charge on the solute, $2.3RT/F$ has a value of about 60 mV at 37°C, and $[\mathrm{So}]_1$ and $[\mathrm{So}]_2$ refer to the concentrations of solute on either side of the membrane. Consider a planar phospholipid bilayer separating two compartments of equal volume. Side 1 contains 50 mM KCl and 50 mM NaCl, whereas side 2 contains 100 mM KCl.

a. If the membrane is made permeable only to K^+, for example, by addition of valinomycin, what is the magnitude of $\Delta\Psi$?
b. If the membrane is made permeable to H^+ and K^+, in which direction does H^+ initially flow?
c. If the membrane can be made selectively permeable to both K^+ and Cl^-, what is the value of $\Delta\Psi$ and the ion concentrations on both sides of the membrane at equilibrium? (*Hint:* Initially, K^+ diffuses down its concentration gradient accompanied by an equivalent amount of Cl^-. Equilibrium is established when the potentials due to K^+ and Cl^- equal each other and the overall membrane potential.)

14. Predict the effects of the following on the initial rate of glucose transport into vesicles derived from animal cells that accumulate this sugar by means of Na^+ symport. Assume that initially $\Delta\Psi = 0$, $\Delta pH = 0$ (pH = 7), and the outside medium contains 0.2 M Na^+, whereas the vesicle interior contains an equivalent amount of K^+.
   a. Valinomycin.
   b. Gramicidin A.
   c. Nigericin.
   d. Preparing the membrane vesicles at pH 5 (in 0.2 M KCl), resuspending them at pH 7 (in 0.2 M NaCl), and adding 2,4-dinitrophenol.

15. In *E. coli*, lactose is taken up by means of proton symport, maltose by means of a binding (ABC-type) protein-dependent system, melibiose by means of Na^+ symport, and glucose by means of the phosphotransferase system (PTS). Although this bacterium normally does not transport sucrose, suppose that you isolated a strain that does. How do you determine whether one of the four mechanisms just listed is responsible for sucrose transport in this mutant strain?

16. You are growing some mammalian cells in culture and measure the uptake of D-glucose and L-glucose (see data in the chart). What type of transport is observed with these sugars? (*Hint:* Plot *V* versus [sugar] and $1/V$ versus 1/[sugar].) Explain the significance of these data.

	V(mM·cm s^{-1}) × 10^7	
[Sugar](mM)	***D-glucose***	***L-glucose***
0.100	166	4.8
0.167	252	8.0
0.333	408	16
1.000	717	50

Solutions

1. The major ingredients are vegetable oil, water and vinegar. The lecithin serves as an emulsifying agent that allows the aqueous and lipid phases to be dispersed in each other and increases the time required for phase separation.

3. The fatty acyl groups are placed in these two different positions by different enzymes (see fig. 19.1). These two enzymes obviously have different specifications for fatty acyl groups.
5. Globular proteins found in the cytosol have tertiary structures that allow the hydrophobic amino acid side chains to assume a more thermodynamically favorable association in the interior of the protein because water is excluded. The polar side chains are most frequently found on the surface of the protein in contact with the aqueous environment. Hydration of the polar groups and the diminished organization of water upon removal of the nonpolar side chains from aqueous contact stabilize the water-soluble protein.

   Integral membrane proteins have a large portion of the surface "dissolved " in the hydrophobic lipid bilayer of the membrane. In this case, exterior polar groups would be less stable then exterior hydrophobic groups.

   The hydrophobic amino acid side chains on the exterior of the integral membrane protein are stable in the water-free environment of the hydrophobic membrane interior. The polar side chains must be inside the protein, associated with water or in ionic bonds with other polar side chains. The portion of the membrane protein protruding from the membrane and in contact with the aqueous environment is composed primarily of polar amino acids, similar to soluble proteins.
7. Peripheral proteins are bound to the inner or outer aspects of the membrane through weak ionic interactions that include association with phospholipid head groups, by electrostatic or ionic interaction with a hydrophilic region of an integral membrane protein or through divalent metal ion bridging to the membrane surface. Integral proteins are dissolved into the lipid bilayer of the membrane through interactions of the hydrophobic amino acid side chains and fatty acyl groups of phospholipids. These interactions exclude the hydrophobic residues from aqueous contact.

   Peripherally bound proteins may be released without disrupting the membrane. Thus, increased salt concentration shields ionic charges and weakens the charge—charge interactions between the peripheral protein and the membrane components. Although increased ionic strength may weaken the salt bridges contributed by divalent metal ions, application of a chelator to sequester the divalent ions may promote release of the peripheral protein. Changing the relative proportion of protonated/deprotonated groups by adjusting pH would, in principle, affect binding of peripheral proteins to the membrane.

   None of the conditions described would be expected to release integral membrane proteins. For example, high ionic strength fosters, rather than weakens, hydrophobic interactions that bind integral proteins to the membrane. In order to remove integral membrane proteins, the membrane must be disrupted by addition of detergents or other chaotropic reagents to solubilize the protein and to prevent aggregation and precipitation of the hydrophobic proteins upon their removal from the membrane. For example, succinate dehydrogenase, a membrane-bound primary dehydrogenase, is an integral protein in the inner mitochondrial membrane and is removed only upon dissolution of the membrane with chaotropic reagents.
9. The amino acid side chains in an α-helix protrude from the axis of the helix and interact either with solvent or with other amino acid side chains (in a folded protein). In the α-helix of single-span integral protein, each amino acid side chain would interact with the hydrophobic interior of the membrane. In principle, it is energetically unfavorable to place a hydrophilic residue in a nonaqueous environment. In integral proteins with multiple α helices that span the membrane, hydrophilic side chains from different helical segments may interact and in some cases form a channel through which ions may diffuse. Portions of the helical segments exposed to the lipid will contain primarily hydrophobic amino acid residues.

11. It is difficult to visualize a protein actually revolving in a controlled manner within the membrane. Most individuals visualize transport proteins creating a channel (conduit, tube, tunnel) through which the material being transported can pass through the membrane. See figure 17.29 for the switching gate idea.

13. a.

$$\Delta\Psi = -2.3\ \frac{RT}{F}\ \log \frac{[K^+]_1}{[K^+]_2} = -60\ \log \frac{50}{100}$$

$$= -60(-0.3) = +18\ \text{mV}$$

b. From side 1 to side 2 because the potential due to K^+ diffusion makes side 1 positive relative to side 2, and H^+ will flow down this electrical gradient.

c. Let X equal the millimolar concentrations of K^+ and Cl^- that are lost from side 2 and gained by side 1. At equilibrium,

$$-2.3\ \frac{RT}{F}\ \log \frac{[K^+]_1}{[K^+]_2} = +2.3\ \frac{RT}{F}\ \log \frac{[Cl^-]_1}{[Cl^-]_2}$$

Therefore,

$$\frac{[K^+]_2}{[K^+]_1} = \frac{[Cl^-]_1}{[Cl^-]_2} \quad \text{or}$$

$$\frac{(100 - X)}{(50 + X)} = \frac{(100 + X)}{(100 - X)}$$

Solving for X gives 14.3 mM. Therefore, at equilibrium,

Side 1: 64.3 mM K^+, 50 mM Na^+, 144.3 mM Cl^-
Side 2: 85.7 mM K^+, 85.7 mM Cl^-

To calculate $\Delta\Psi$, one can use either the K^+ or Cl^- gradient:

$$\Delta\Psi = -60\ \log \frac{64.3}{85.7} = +60\ \log \frac{114.3}{85.7}$$

$$= 60(0.125) = 7.5\ \text{mV}$$

(This potential is such that side 1 is positive and side 2 is negative.)

15. This problem could be approached in a number of ways, but the simplest of these will be described for illustrative purposes.

a. If sucrose transport were energized via H^+ symport, membrane vesicles should transport sucrose in the presence of an electron donor, such as D-lactate, which would set up a ΔP. Proton ionophores should inhibit sucrose uptake under these conditions.

b. If Na^+ symport were involved, active uptake should be absolutely dependent on extravesicular (or extracellular) Na^+, stimulated by a $\Delta\Psi$ (interior negative) and insensitive to ΔpH (at constant $\Delta\Psi$).
c. If there were a sucrose binding protein necessary for transport, cells subjected to cold osmotic shock, spheroplasts, and plasma membrane vesicles should all be defective in sucrose transport regardless of the energy source tested.
d. If a PTS for sucrose were present, crude extracts of cells should phosphorylate sucrose by a reaction dependent on PEP, but not on ATP.

18 Metabolism of Fatty Acids

Summary

In this chapter we focused on the synthesis and degradation of long-chain fatty acids and on how these processes are regulated. Most of our discussion was concerned with how these reactions take place in the mammalian liver, although occasionally we referred to other animal tissues and to *E. coli*. The following points are the most important.

1. Fatty acids originate from three sources: diet, adipocytes, and *de novo* synthesis.
2. The degradation of fatty acids occurs by an oxidation process in the mitochondria. The breakdown of the 16-carbon saturated fatty acid, palmitate, occurs in blocks of two carbon atoms by a cyclical process. The active substrate is the acyl-CoA derivative of the fatty acid. Each cycle involves four discrete enzymatic steps. In the process of oxidation the energy is sequestered in the form of reduced coenzymes of FAD and NAD^+. These reduced coenzymes lead to ATP production through the respiratory chain. The oxidation of fatty acids yields more energy per carbon than the oxidation of glucose because saturated fatty acids are in the fully reduced state.
3. Unsaturated fatty acids are also oxidized in mitochondria with the help of certain additional enzymes that facilitate a continuous flow of the oxidation process.
4. The main end product of fatty acid oxidation is acetyl-CoA, which can be used by the tricarboxylic acid cycle for generation of energy. Alternatively, ketone bodies may be formed from condensation of acetyl-CoAs. Ketone bodies are made in the liver and subsequently diffuse into the blood to be carried to other tissues, where they are converted into acetyl-CoA for various metabolic purposes.
5. Fatty acid biosynthesis also occurs in steps of two carbon atoms. Biosynthesis takes place in the cytosol. In addition to occurring in a different cellular compartment from degradation, biosynthesis involves totally different enzymes and different coenzymes.
6. Fatty acid synthesis takes place in eight steps. All except the first step take place on a multienzyme complex. The intermediates on this complex are carried by attachment of the acid group in thioester linkage to phosphopantetheine of the acyl carrier protein (ACP). The multienzyme complex greatly increases the efficiency of fatty acid synthesis, because for each step in the pathway the next enzyme is always near at hand, and the dilution of intermediates is minimized.
7. Regulation of fatty acid metabolism takes place in such a way that simultaneous synthesis and degradation are minimized. Control factors ensure that synthesis occurs primarily when energy is in excess, and degradation when energy is needed. When energy is abundant, the synthesis of malonyl-CoA by acetyl-CoA carboxylase stimulates fatty acid synthesis and inhibits carnitine acyltransferase I and, as a result, β oxidation. The hormones epinephrine and glucagon stimulate degradation and inhibit synthesis. They stimulate degradation by facilitating release of fatty acids stored in the adipocytes. They inhibit synthesis by inactivating the first enzyme in the pathway for synthesis, acetyl-CoA carboxylase. Most other control factors interfere with the supply of substrate for either of the two processes.

Problems

1. The consensus is that fats containing unsaturated fatty acids are "better for you" than the corresponding saturated forms. Can this statement be explained by the ATP yield that results on complete oxidation (which in turn reflects the caloric content)? Calculate the number of ATPs produced for the complete oxidation of arachidic ($C_{20:0}$) and arachidonic ($C_{20:4}$) acids to assess any differences in energy value of saturated versus polyunsaturated fatty acids.

2. Examine the chemistry of the reactions presented in figure 18.4a. Where else in metabolism have you seen a similar sequence of chemical events? What feature(s) is identical, what different?

3. a. Calculate the number of moles of ATP produced during the catabolism of a mole of glucose.
   b. Calculate the number of moles of ATP produced during the catabolism of a mole of decanoic acid [$CH_3(CH_2)_8CO_2H$]. Although this is not a typical fatty acid, it was selected because it has a molecular mass comparable to that of glucose.
   c. If the heat of combustion of glucose is –669.9 kcal/mole and decanoic acid is –1452 kcal/mole, and if –7.3 kcal are preserved per mole of ATP, what fraction (%) of the energy available is preserved in each case?
   d. Consider glucose and decanoic acid to represent a typical carbohydrate and fat, respectively. A typical carbohydrate contains 4 kcal/g, and fats contain 9 kcal/g. Calculate the ratio of the nutritional calories for fats to carbohydrates and compare this number with the ratio of the ATPs produced for fats to carbohydrates. Are the numbers consistent?

4. Order the following substances from the least to most oxidized carbon: Formaldehyde, carbon dioxide, methane, formic acid, methanol. Use the open-chain form of glucose and the structural formula for decanoic acid and this scheme to determine the "average" oxidation state of the carbons in glucose and decanoic acid. How does this oxidation state compare with the data obtained in problems 3a and 3b above? Can you explain your observation?

5. a. Consider the complete oxidation of glucose (M_r = 180) via glycolysis and the TCA cycle, and calculate the moles ATP generated during the oxidation of 1 mole of glucose to CO_2 and H_2O. Assume that the free energy of ATP hydrolysis under physiological conditions is –11 kcal/mole, and assume mitochondrial P/O ratios of 2.5 for NADH oxidation and 1.5 for succinate (or equivalent) oxidation. Estimate the free energy conserved as ATP during the oxidation. Calculate the free energy conserved per gram of glucose oxidized.
   b. Repeat the calculations for ATP formation, but consider the complete oxidation of 1 mole of palmitic acid (M_r = 256) via β oxidation and the TCA cycle. Estimate the free energy conserved as ATP energy from palmitate oxidation. Calculate the free energy conserved per gram of palmitic acid oxidized.
   c. Bearing in mind that respiratory metabolism is an oxidative process, how do you explain the differences in energy content of fat and carbohydrate on a weight basis?
   d. Explain the rationale behind the use of fat rather than carbohydrate as an energy reserve in plants and animals.

6. Explain the role of carnitine acyltransferases in fatty acid oxidation.

7. Late-night TV ads are often quite educational. One advertisement touted the merits of eating grapefruit to lose weight, but taking them in your lunch presents problems. At that point an individual in a lab coat standing in front of a blackboard covered with molecular formulas proceeds to explain how it is the citric acid in the grapefruit that burns up the fat and for only $XX you can get a month's supply of their pills. Use what you know about the citric acid cycle and fatty acid metabolism to deduce if added citric acid causes weight loss.

8. After working through problem 7 and having read the chapter, you want to cash in on a get-rich-quick diet scam. You start thinking, if only the body could be tricked into converting some of its fatty acids into acetyl-CoA and then resynthesize fatty acids, there would be a weight loss. If you could actually make this work, would there be a weight loss?

9. Is β oxidation (fig. 18.4) best described as a spiral or cyclic process? Why?

10. Acetoacetate (fig. 18.7) is shown to give rise to acetone by a spontaneous reaction. Can you explain how this might occur?

11. β-Oxoacyl-CoA transferase (see fig. 18.8) is involved in the transfer of a CoASH from succinyl-CoA to acetoacetate to produce succinate and acetoacetyl-CoA. A cursory examination of this reaction suggests a simple transfer of the CoA moiety. However, it is soon realized that the loss of an oxygen by the acetoacetate and the gain of an oxygen by the succinyl group present a dilemma. Produce a rational mechanism that explains the preservation of the thioester energy and solves this dilemma. *Hint:* Consider a succinyl phosphate intermediate.

12. Carnitine deficiency in liver is correlated with hypoglycemia. Suggest a plausible explanation for hypoglycemia in the carnitine-deficient human.

13. a. Liver mitochondria convert long-chain fatty acids to ketone bodies (acetoacetate and β-hydroxybutyrate) that are subsequently transported in the plasma to nonhepatic tissues. Suggest some metabolic advantages of supplying ketone bodies to nonhepatic tissues.
    b. In what way is β-hydroxybutyrate a better energy source than acetoacetate for nonhepatic tissues?
    c. Outline the oxidation of β-hydroxybutyrate to acetyl-CoA in heart mitochondria.

14. Predict the effect on oxidation of ketone bodies and of glucose in nonhepatic tissue of individuals with markedly diminished β-oxyacid-CoA-transferase activity. Predict the effect if the activity was absent.

15. a. For an *in vitro* synthesis of fatty acids with purified fatty acid synthase, the acetyl-CoA was supplied as the ^{14}C-labeled derivative.

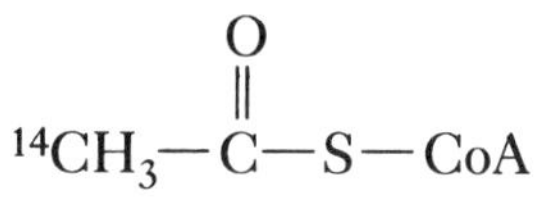

$$^{14}CH_3-\overset{\overset{\displaystyle O}{\|}}{C}-S-CoA$$

The other reactants, including the malonyl-CoA, were not radioactive. Where is the ^{14}C-label found in palmitic acid?

b. If the malonyl-CoA was supplied as the only labeled compound deuterated as shown in the structure below, how many deuterium atoms would be incorporated in palmitate? On which carbon(s) would these deuterium atoms reside?

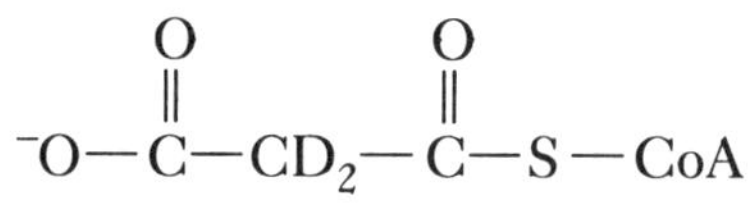

$$^{-}O-\overset{\overset{\displaystyle O}{\|}}{C}-CD_2-\overset{\overset{\displaystyle O}{\|}}{C}-S-CoA$$

c. If [3-^{14}C]malonyl-CoA (shown below) was used in the reaction, which atoms in palmitate would be labeled? Why?

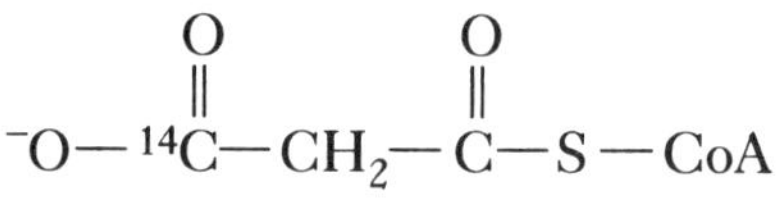

$$^{-}O-{}^{14}\overset{\overset{\displaystyle O}{\|}}{C}-CH_2-\overset{\overset{\displaystyle O}{\|}}{C}-S-CoA$$

16. Except for malonyl-CoA formation, all the individual reactions for palmitate synthesis reside on a single multifunctional protein (fatty acid synthase) in animal cells. It has been shown that a dimer of the multifunctional protein is required to catalyze palmitate synthesis. Explain the molecular basis of this observation.

17. What are the metabolic sources of NADPH used in fatty acid biosynthesis? How many moles of NADPH are required for the synthesis of 1 mole of palmitic acid from acetyl-CoA?

18. Citrate is both a lipogenic substrate and a regulatory molecule in mammalian fatty acid synthesis.
    a. Explain each function of citrate in fatty acid synthesis.
    b. Write reactions (including structures) outlining the role of citrate as a lipogenic substrate.

19. Which catalytic activity of the mammalian fatty acid synthase determines the chain length of the fatty acid product?

20. a. Why is the location of biosynthesis and β oxidation of fatty acids in separate metabolic compartments essential to regulation of fatty acid metabolism in the hepatocyte?
    b. Would you expect an inhibitor of the extramitochondrial carnitine acyltransferase to mimic the effect of malonyl-CoA on β oxidation? (Assume that the inhibitor can penetrate the cell membrane.) Explain the rationale for your answer.

Solutions

1. Using calculations based on figures 18.4 and 18.6 and assuming NADPH = NADH, arachidic produces 134 moles of ATP/mole FA while arachidonic gives 126 moles of ATP/mole of fatty acid. This difference in ATPs is rather minor. It is unlikely, therefore, that any biological difference caused by dietary saturated vs. unsaturated fats is due to their ATP yields.

Arachidic acid		ATP's
beta-oxidation produces 10 AcSCoA	10×10	=100
9 $FADH_2$ + 9 NADH	9×4	= 36
fatty acid activation		−2
sum		134

Arachidonic acid (see figure 18.6)		
beta-oxidation produces 10 AcSCoA	10×10	=100
7 $FADH_2$ + 9 NADH − 2 NADPH =	$7 \times 1.5 + 7 \times 2.5$	= 28
fatty acid activation		−2
sum		126

3. a. Glucose

2 cytoplasmic NADH	3
2 cytoplasmic ATP	2
2 NADH from pyr deH_2ase	5
2 AcSCoA	20
sum	30 ATP

b. 64 moles ATP per mole decanoic acid

c.

30 ATP	−7.5 Kcal	Glucose	100	
glucose	ATP	−669.3 Kcal		= 33%

64 ATP	−7.5 Kcal	C10 Acid	100	
C10 acid	ATP	−1452 Kcal		= 32%

d. $\dfrac{9 \text{ kcal/g fat}}{4 \text{ kcal/g carbo}} = 2.25 \qquad \dfrac{64 \text{ ATP/fat}}{30 \text{ ATP/glu}} = 2.13$

The ratios are quite close.

5. a. Glycolytic oxidation of 1 mole glucose to pyruvate yields 2 moles pyruvate, 2 moles NADH, and 2 moles ATP. The extramitochondrial NADH is assumed to reoxidized via the malate shuttle, yielding approximately 2.5 moles ATP. The mitochondrial oxidation of pyruvate to CO_2 plus H_2O yields 12.5 moles ATP each. The total ATP generation is 32 moles ATP per mole of glucose. Total energy conserved as ATP can be estimated: 32 moles ATP × 11 kcal/mole = 350 kcal/mole. Energy conserved per gram of glucose is 350 kcal/mole/180/gm/mole = 1.9 kcal/gm^{-1}.

b. Oxidation of palmitic acid requires the activation of the acid and formation of a thioester with coenzyme A. The synthase uses the equivalent of 2 moles ATP (hydrolysis of two phosphoanhydride bonds) in the reaction

$$RCOO^- + ATP + CoASH \rightarrow RCO\text{–}SCoA + AMP + \text{pyrophosphate}$$

Pyrophosphatase hydrolyzes the pyrophosphate. Thus, mole AMP plus 2 moles P_i may be considered end products.

β-Oxidation of 1 mole palmitoyl-CoA yields 8 moles acetyl-CoA in seven repetitions of the cycle. Seven moles FADH and 7 moles NADH are produced in the mitochondrial matrix. Transfer of these reducing equivalents to O_2 yields 28 moles ATP. Oxidation of the 8 moles acetyl-CoA in the TCA cycle yields 8 × 10 moles ATP. The total ATP yield from *β* oxidation plus the TCA cycle yields 108 moles ATP, but 2 moles ATP were used to activate the fatty acid. Net yield is 106 moles ATP per mole of palmitic acid oxidized. The energy stored as ATP is 106 moles ATP per mole of palmitic acid × 11 kcal per mole ATP = 1,170 kcal/mole conserved. Energy conserved per gram is 1,170 kcal/mole/256 gm/mole = 4.6 kcal/gm.

c. Lipids are more highly reduced than are carbohydrates (palmitic acid: $C_{16}H_{32}O_2$; glucose: $C_6H_{12}O_6$). Removal of reducing equivalents from substrate and subsequent reduction of O_2 releases energy that is conserved by oxidative phosphorylation. Lipids provide more reducing equivalents per gram than do carbohydrates, and they provide more energy for oxidative phosphorylation.

d. Fat has approximately 2.4 times more energy stored per gram than does carbohydrate, and the energy is stored in compact hydrophobic droplets in specialized cells (adipocytes). On that basis, one would have to store 2.4 times the mass of carbohydrate to equal the energy stored by lipid. Carbohydrate mass likely would be even larger because of water of hydration present in the stored form.

7. The availability of citrate has no relationship to the flow of metabolites through the tricarboxylic acid cycle. Increased citrate concentrations result in increased cytoplasmic acetyl-CoA concentrations which in turn increases fatty acid biosynthesis.

9. Viewed from a reaction sequence viewpoint it is a cyclic process. From a substrate viewpoint it is a decreasing spiral because the substrate's length decreases with each turn around the spiral.

11. Many mechanisms are conceivable. One possibility is: 1) approach of a phosphate on succinyl-CoA to produce succinyl phosphate (a mixed anhydride), 2) approach of an acetoacetate oxygen on the mixed anhydride to transfer a phosphoryl (PO_3) group and produce acetoacetyl phosphate, 3) approach by CoASH on the C-1 carbonyl of acetoacetyl phosphate to eliminate a phosphate and produce acetoacetyl-CoA.

$^-O_2CCH_2CH_2-C(=O)-S-CoA$ + P_i → (CoASH) → $^-O_2CCH_2CH_2-C(=O)-O-P(=O)(O^-)-O^-$

$CH_3-C(=O)-CH_2-C(=O)-O^-$

→ succinate

$CH_3C(=O)CH_2C(=O)-S-CoA$ ← (P_i) ← $CH_3-C(=O)-CH_2-C(=O)-O-P(=O)(O^-)-O^-$ + CoASH

13. a. Nonhepatic tissues use ketone bodies as a rich energy source supplied by the liver. The ketone bodies are water soluble and easily transported in the blood. Oxidation of ketone bodies by heart muscle, for example, spares the oxidation of glucose, allowing the carbohydrate to be metabolized by erythrocytes and the brain. Oxidation of 1 mole acetoacetate yields a net 19 moles ATP.

b. β-Hydroxybutyrate supplies an additional reducing equivalent per mole compared to acetoacetate. Oxidation of 1 mole β-hydroxybutyrate yields 1 mole acetoacetate plus 1 mole NADH. Subsequently, oxidation of the NADH provides 2.5 additional moles ATP compared to the oxidation of acetoacetate.

c.

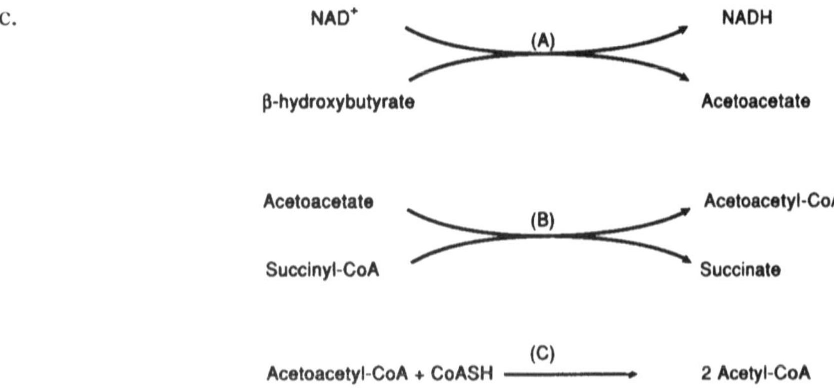

(A): NAD^+-dependent β-hydroxybutyrate dehydrogenase
(B): 3-ketoacyl-CoA transferase
(C): thiolase

15. a. Acetyl-CoA is used directly only in the first cycle of condensation and reduction in palmitate synthesis. The methyl group of acetyl-CoA will become C-16 of palmitate. If 2-^{14}C acetyl-CoA were the only labeled metabolite, only C-16 of palmitate would be labeled.
 b. Synthesis of palmitate requires condensation and reduction of one acetyl and seven malonyl units. Each malonyl-CoA unit added undergoes two reduction steps with an intervening dehydration. The dehydration step removes one of the two protons from the methylene group derived from malaonyl-CoA. Thus the product will retain only one of the two deuterium labels as shown.

$$CH_3{-}\overset{O}{\overset{\|}{C}}{-}CD_2{-}\overset{O}{\overset{\|}{C}}{-}S{-}ACP \xrightarrow[+H^+]{NADPH\ \ NADP^+} CH_3{-}\overset{OH}{\overset{|}{\underset{H}{\underset{|}{C}}}}{-}CD_2{-}\overset{O}{\overset{\|}{C}}{-}S{-}ACP$$

$$CH_3{-}\overset{OH}{\overset{|}{C}}H{-}CD_2{-}\overset{O}{\overset{\|}{C}}{-}S{-}ACP \xrightarrow{-D^+,\ OH^-} CH_3{-}CH{=}CD{-}\overset{O}{\overset{\|}{C}}{-}S{-}ACP$$

$$CH_3{-}CH{=}CD{-}\overset{O}{\overset{\|}{C}}{-}S{-}ACP \xrightarrow[+H^+]{NADPH\ \ NADP^+} CH_3{-}CH_2{-}CHD{-}\overset{O}{\overset{\|}{C}}{-}S{-}ACP$$

There will be seven deuterium atoms per palmitate molecule, one residing on each of the carbons 2, 4, 6, 8, 10, 12, and 14. Carbon 16 originates directly from acetyl-CoA and is not labeled.
 c. Condensation of two-carbon units to the growing fatty acyl chain is driven in part by decarboxylation of the malonyl unit. The carbon shown labeled in the problem is lost as CO_2 during decarboxylation and is not incorporated into palmitate.

$$CH_3{-}\overset{O}{\overset{\|}{C}}{-}S{-}\text{acyl carrier protein}$$

$$\text{--- --- --}\!\!\blacktriangleright \quad CH_3{-}\overset{O}{\overset{\|}{C}}{-}CH_2{-}\overset{O}{\overset{\|}{C}}{-}S{-}ACP$$

$$^{-}O{-}^{14}C{-}CH_2{-}\overset{O}{\overset{\|}{C}}{-}S{-}ACP \qquad {}^{14}CO_2$$

$$+\ ACP{-}SH$$

17. NADPH for fatty acid biosynthesis is supplied by the glucose-6-phosphate and 6-phosphogluconate dehydrogenases in the pentose pathway and by the NADP$^+$-dependent-malic enzyme. Oxaloacetate released by the ATP-citrate lyase may be reduced to L-malate by the cytoplasmic NAD$^+$-dependent malate dehydrogenase. Malic enzyme oxidatively decarboxylates L-malate to pyruvate and reduces NADP$^+$ to NADPH. Pyruvate enters the mitochondrial matrix for resynthesis of oxaloacetate by carboxylation.

Palmitate formation requires seven cycles of malonyl-CoA condensation, each of which requires 1 mole NADPH for reduction of the ketoacyl group to β-hydroxyacyl-ACP and another to reduce the enoyl-ACP. Reduction of 1 mole of acetyl-CoA and 7 moles of malonyl-CoA to form 1 mole of palmitate requires 14 moles NADPH.
19. Thiolesterase activity of the fatty acid synthase is more active with palmityl-ACP than with shorter fatty acyl ACPs. Palmitate, the 16-carbon fatty acid, is thus preferentially released.

19 Biosynthesis of Membrane Lipids

Summary

Fatty acids are components of a rich variety of complex lipid molecules that play critical structural roles in membranes. Some of these lipids are also the precursors of compounds with hormone or second-messenger activities. In this chapter we focused on the metabolism of these compounds with some mention of their functions. The following points are the highlights of our discussion.

1. Lipid synthesis is unique in that it is almost exclusively localized to the surface of membrane structures. The reason for this restriction is the amphipathic nature of the lipid molecules. Phospholipids are biosynthesized by acylation of either glycerol-3-phosphate or dihydroxyacetone phosphate to form phosphatidic acid. This central intermediate can be converted into phospholipids by two different pathways. In one of these, phosphatidic acid reacts with CTP to yield CDP-diacylglycerol, which in bacteria is converted to phosphatidylserine, phosphatidylglycerol, or diphosphatidylglycerol. In *E. coli*, the major phospholipid, phosphatidylethanolamine, is synthesized by means of this route through the decarboxylation of phosphatidylserine. In the second pathway, found in eukaryotes, phosphatidic acid is hydrolyzed to diacylglycerol, which reacts with CDP-ethanolamine or CDP-choline to yield phosphatidylethanolamine or phosphatidylcholine, respectively. Alternatively, the diacylglycerol may react with acyl-CoA to form triacylglycerol.
2. There are numerous reactions by which the acyl groups or polar head-groups of phospholipids might be modified or exchanged. Phospholipids are degraded by specific phospholipases.
3. In *E. coli*, phospholipids are used almost exclusively as structural components of the cell membranes, and regulation is known to occur at an early stage in fatty acid synthesis. In the mammalian liver, fatty acids are important precursors of both the structural phospholipids and also of the energy-storage lipid, triacylglycerol. The cell's need for structural lipids is satisfied before fatty acids are shunted into energy storage. The rate of triacylglycerol synthesis is regulated by the availability of fatty acid from both diet and biosynthesis. The regulation of phosphatidylcholine biosynthesis occurs primarily at the reaction catalyzed by CTP: phosphocholine cytidylyltransferase. This enzyme is activated by translocation from a soluble form, where it is inactive, to cellular membranes, where it is activated. The regulation of the biosynthesis of other phospholipids in eukaryotic cells is less well understood.
4. The sphingolipids are important structural lipids found in eukaryotic membranes. The acylation of sphingenine produces ceramide, which reacts with phosphatidylcholine to give sphingomyelin. Ceramide also can react with activated carbohydrates (e.g., UDP-glucose) to form the glycosphingolipids. Studies on the catabolism of the sphingolipids have revolved around inherited diseases, sphingolipidoses, that result from a defect in lysosomal enzymes that degrade sphingolipids.
5. Prostaglandins and thromboxanes are hormonelike substances that affect the function of a cell by binding to a receptor on the cell surface. Prostaglandins and thromboxanes are biosynthesized from C_{20} polyunsaturated fatty acids, primarily arachidonic acid. In the initial reaction, arachidonic acid is converted to PGH_2 by prostaglandin endoperoxide synthase. PGH_2 is then converted to PGE_2, PGF_2, TXA_2, and PGI_2.

Problems

1. Would you expect phosphatidylserine decarboxylase (fig. 19.3) to be a pyridoxal phosphate enzyme?

2. Phosphatidylcholine biosynthesis appears to be regulated principally at the step catalyzed by CTP: phosphocholine cytidylyltransferase. Does the type of regulation observed make biochemical sense? Draw a chemical reaction mechanism for this enzyme.

3. Explain why the reactions from choline to phosphatidylcholine in eukaryotic cells are thermodynamically feasible.

4. A patient accumulated the lipid Gal-β(1,4)-Glc-β(1,4)-ceramide. What enzymatic reaction might be defective?

5. Sphingolipids as well as other membrane components are constantly degraded and resynthesized. Why does the cell waste so much energy for what appears to be a futile effort?

6. If both parents carry a single defective gene for the hexosaminidase A enzyme, there is a 25% chance that their child will have Tay-Sachs disease. What chance is there that their child would be a carrier of a defective gene?

7. Why don't carriers of a defective gene for the hexosaminidase A enzyme suffer from Tay-Sachs disease (or at least some of the symptoms)?

8. One of the pathways for the biosynthesis of phosphatidic acid in eukaryotes originates with dihydroxyacetone phosphate (from glycolysis). This pathway is shown only as three unlabeled arrows in figure 19.2. Given that the three enzymes starting with dihydroxyacetone phosphate are (1) an acyl transferase, (2) a dehydrogenase, and (3) a second acyltransferase, complete the pathway including any missing substrates.

9. This chapter discusses many of the biosynthetic interconversions that are known to occur with the different phosphatidyl components (e.g., phosphatidylethanolamine, phosphatidylcholine, and phosphatidylinositol) of membranes. Only considering the bioenergetics, what is the major difference between the biosynthetic pathways shown in figures 19.4 and 19.6?

10. Low doses of aspirin (one aspirin every other day) are recommended to prevent heart attacks and strokes. Why do three or four tablets per day not work better? (*Hint:* Remember that TXA_2 is made in platelets and that PGI_2 is made in the arterial walls.)

11. Snake venoms contain many types of lipases, including phospholipase A_2. Why do small amounts of this enzyme contribute to some of the toxic effects of snake venom? (Bee venom contains a protein that stimulates phospholipase A_2.)

Solutions

1. Phosphatidylserine decarboxylase is a pyridoxal phosphate enzyme. This is expected because phosphatidylserine has a free amino group and typically amino acid decarboxylases are pyridoxal phosphate enzymes.

3. The synthesis of phosphatidylcholine is thermodynamically feasible because of the ATP used to phosphorylate choline and the CTP used to form CDP-choline. The cytidylyltransferase reacts phosphocholine with CTP to form CDP-choline and pyrophosphate. The hydrolysis of this pyrophosphate by pyrophosphatase is the major driving force for these reactions.
5. One possibility is the removal of fatty acyl groups that have been damaged by oxidation with molecular oxygen (or other oxygen species). In addition these processes allow the adjustment of the fluidity of the membrane to any environmental changes.
7. Carriers of a defective gene for the hexosaminidase A enzyme still have a functional copy of the gene. In this case, one functional gene is capable of producing enzyme amounts sufficient to prevent the disease. This produces a situation known as a recessive disease.
9. Figure 19.4 involves a CDP-ethanolamine or CDP-choline which reacts with diacylglycerol to produce phosphatidyl-ethanolamine or phosphatidylcholine. Figure 19.6 has a CDP-diacylglycerol that reacts with an inositol to produce a phosphatidylinositol. The activated component is different, in the first case (fig. 19.4) the minor component is activated while in the production of phosphatidylinositol (fig. 19.6) the diacylglycerol is activated.
11. Phospholipase A_2 selectively removes fatty acids from the SN-2 position of phospholipids. Arachidonic acid is stored in membranes by linkage at the SN-2 position in phospholipids. If a lot of arachidonic acid were released, it would stimulate the synthesis of prostaglandins, which then would induce inflammation. The release of arachidonic acid from phospholipids is believed to be the rate-limiting step in eicosanoid biosynthesis.

20 Metabolism of Cholesterol

Summary

In this chapter we dealt primarily with the metabolism of cholesterol, the most prominent member of the steroid family of lipids, and with the associated plasma lipoproteins. The chief points in our discussion are as follows:

1. The biosynthesis of steroids begins with the conversion of three molecules of acetyl-CoA into mevalonate, the decarboxylation of mevalonate, and its conversion to isopentenyl pyrophosphate. Six molecules of isopentenyl pyrophosphate are polymerized into squalene, which is cyclized to yield lanosterol. Lanosterol is converted to cholesterol, which is the precursor of bile acids and steroid hormones.
2. The rate of cholesterol biosynthesis appears to be regulated primarily by the activity of HMG-CoA reductase. This key enzyme is controlled by the rate of enzyme synthesis and degradation and by phosphorylation-dephosphorylation reactions. Synthesis of the mRNA for the reductase is inhibited by cholesterol delivered to cells by means of low-density lipoproteins (LDLs).
3. Cholesterol, triacylglycerols, and phospholipids are carried in plasma by lipoproteins, which are synthesized and secreted by the intestine and liver. The major lipoproteins are chylomicrons, very-low-density lipoproteins (VLDLs), low-density lipoproteins (LDLs), and high-density lipoproteins (HDLs). The triacylglycerols in chylomicrons and VLDLs are degraded in plasma by lipoprotein lipase, and the fatty acids are absorbed primarily by heart, skeletal muscle, and adipose tissue. LDLs are removed from plasma by an endocytotic process after binding to specific LDL receptors on the plasma membrane. The LDLs are enzymatically degraded in the lysosomes. In familial hypercholesterolemia, the specific receptors for LDL uptake are defective. High levels of LDL are associated with an increased risk of cardiovascular disease, whereas high levels of HDL seem to protect against this disease.
4. Bile acids are C_{24} acids that are biosynthetically derived from cholesterol. The 7α-hydroxylation of cholesterol is the committed and rate-limiting reaction in the synthesis of bile acids. Salts formed from the bile acids are secreted into the small intestine and aid the solubilization and digestion of lipids. Formation and excretion of the bile salts is the major route for elimination of cholesterol from the body.
5. Steroid hormones are biosynthesized from cholesterol in the adrenal cortex, gonads, and placenta. These steroids are important hormones for many specific physiological processes.

Problems

1. Cholyl-CoA to glycocholate (fig. 20.21) is an example of amide bond formation utilizing a CoASH (thioester) derivative. Are all amide bonds made from CoASH derivatives?

2. The conversion of cholic acid to cholyl-CoA (fig. 20.21) also involves the conversion of ATP to AMP and PP_i. What intermediate would you expect to be involved in this reaction?

3. Explain the frequencies of the heterozygous and homozygous forms of a familial hypercholesterolemia given in the text (page 472). Do you find anything inconsistent with these frequencies? If so, can you provide an explanation for this?

4. Do you expect the equilibrium position of the reaction catalyzed by CoA: cholesterol acyltransferase (fig. 20.13) to be any different from that of lecithin: cholesterol acyltransferase (fig. 20.17)?

5. A patient homozygous for familial hypercholesterolemia (FH) was treated with lovastatin to lower LDL levels in the blood. This treatment did not have any effect on LDL levels. Why? After a number of heart attacks, a heart and liver transplant were done, and LDL levels were dramatically lowered. Why were both organs replaced?

6. During routine investigations, the plasma from a family of rats (group 1) was found to have very low concentrations of cholesterol. When the microsomal HMG-CoA reductase from liver was assayed, extremely low activities were found. When the cytosol from normal rats (group 2) was added to the microsomal fraction from group 1 rats, the HMG-CoA reductase activity was gradually restored to normal values. What enzyme activity (or activities) might be deficient in the group 1 rats?

7. Liver cells in culture are given 2-[^{14}C]-acetate. Where does this label appear in HMG-CoA?

8. How would a dietary resin that absorbs bile salts reduce plasma cholesterol levels?

9. Notice that the metabolic sequences described in this and previous chapters often involve multiple locations. These locations can be at the organ (fig. 20.1) or the subcellular level (fig. 20.22). This is in contrast, for example, with glycolysis, which occurs in the cytosol, or the tricarboxylic acid cycle, which is completely mitochondrial. Can you provide a possible explanation for this multiple location phenomenon?

Solutions

1. No. Some examples of other types of intermediates involved in amide bond formation include: a) mixed anhydrides, in peptidoglycans (fig. 16.16), glutamine (fig. 18.3) and citrulline (fig. 19.6) synthesis, b) esters, amino acyl-tRNAs in peptidoglycan synthesis (fig. 16.17) and protein synthesis (chapter 29).
3. These frequencies are consistent if one assumes that the gene has only two alleles: $A + B = 1$ and $A^2 + 2AB + B^2 = 1$ describes the population where A^2 are "normal," 2AB are heterozygous and B^2 are homozygous for familial hypercholesterolemia. $2AB = 1/500$, $AB = 0.001$, $A = 0.999$ and $B = 0.001$. Therefore, $B^2 = (0.001)^2 = 0.000001$, which is the one in a million indicated in the text. These frequencies would be inconsistent if the gene has more than two alleles.
5. People who are homozygous for familial hypercholesterolemia (FH) do not make any receptors for LDL and cannot remove LDL from their blood. This leads to very high levels of serum LDL, which results in cardiovascular disease. (Some of the highest levels of serum cholesterol observed have been found in FH patients.) If a normal person is given an inhibitor for HMG-CoA reductase, cholesterol synthesis is inhibited in the liver. Lower levels of cholesterol then signal the synthesis of increased levels of LDL receptors. This increases the uptake of LDL into the liver and reduces serum LDL. In a patient with FH, this has little effect, since there are no LDL receptors. The only effect is that the liver does not make as much cholesterol and does not contribute as much to serum LDL levels. The patient in question was given a heart transplant to replace the damaged heart and, at the same time, a liver transplant. The new liver will make normal amounts of LDL receptors and have normal uptake of LDL from the blood. This will dramatically lower serum LDL levels and prevent the new heart from developing coronary artery disease. If the liver transplant had not been done, the heart transplant would have been to no avail.

7. The ^{14}C in HMG-CoA derived from 2-[^{14}C]-acetate is marked in the structure.

$$^{-}O-\overset{O}{\overset{\|}{C}}-\overset{*}{C}H_2-\underset{\overset{*}{C}H_3}{\overset{OH}{\overset{|}{C}}}-\overset{*}{C}H_2-\overset{O}{\overset{\|}{C}}-S-CoA$$

$* = {}^{14}C$

9. Essentially any attempt at an explanation requires evolution concepts. The increasing involvement of membranes, organelles, and organs during evolution combined with the evolutionary development of new metabolic sequences from previously developed enzymatic mechanisms has given us what appears to be a diverse collection of chemical components and processes. Notice within this chapter, for example, what diverse roles nature has found for the steroid ring system.

21 Amino Acid Biosynthesis and Nitrogen Fixation in Plants and Microorganisms

Summary

In this chapter we discussed the biosynthesis of amino acids and the roles that certain amino acids play in bringing inorganic nitrogen and sulfur into bioorganic compounds.

1. The pathways to amino acids arise as branchpoints from a few key carbohydrate intermediates in the central metabolic pathways. The common starting point for a branched pathway leading to several amino acids defines a family.
2. Amino acid biosynthesis is best studied in microorganisms in which all 20 of the amino acids most commonly found in proteins are synthesized. In microorganisms both genetic and biochemical techniques can be harnessed to analyze the pathways. Typically, research begins by isolating mutants defective in single steps in a particular biosynthetic pathway and analyzing the consequences of the mutation.
3. The glutamate family contains four amino acids: Glutamate, glutamine, proline, and arginine. Only the synthesis of glutamate and glutamine is discussed in this chapter. In some cell types growing in the presence of a high concentration of ammonia, the amination of α-ketoglutarate occurs directly by free ammonia. For most cells and under most conditions, this amination occurs at the expense of the amide group of glutamine. The amide nitrogen of glutamine must be regenerated by the amidation of glutamate. This reaction is a major route for the incorporation of ammonia into bioorganic compounds, and so it is not surprising that the reaction catalyzed by glutamine synthase is a regulated process, which is most sensitive to the sufficiency of the nitrogen supply of the cell.
4. Ammonia (NH_3) is the form in which nitrogen is incorporated into organic materials. Nitrogen exists in the –3 valence state in NH_3. Nitrogen itself actually passes through various forms and valence states as a result of its interactions with different living forms. The valence states range from –5 in nitrates to –3 in ammonia or organic materials. In the 0 valence state, nitrogen is a gas. The passage of nitrogen from one form to another involves a chain of widely distributed organisms. The biological fixation of gaseous nitrogen by both free-living and symbiotic nitrogen-fixing bacteria is catalyzed by an enzyme complex called nitrogenase. Dinitrogen is bound by this complex as it gradually undergoes reduction to ammonia, one electron at a time.
5. The serine family includes three amino acids: Serine, glycine, and cysteine. In this chapter we focused on the synthesis of cysteine, which funnels sulfur into the biochemical world. The biosynthesis of L-cysteine entails the sulfhydryl transfer to an activated form of serine. Most sulfur in nature exists in the inorganic, highly oxidized form of sulfate ion. This sulfur must be reduced to H_2S before it can be incorporated into amino acids.
6. The aspartate and pyruvate families together contain 11 amino acids. Because of the reactions involved in its synthesis, isoleucine is considered a member of both families. Isoleucine and valine use four enzymes in common in their biosynthetic pathways.
7. Chorismate is a common precursor of the amino acids of the aromatic amino acid family. Tryptophan is synthesized in five steps from chorismate.
8. Histidine is in a family of one. There are nine steps in this pathway which interacts with the purine pathway.

9. There are a large number of amino acids found in different organisms that are not incorporated into proteins. For example, the D-amino acids are commonly found in microbial cell walls and in many peptide antibiotics. In most cases the formation of D-amino acid containing peptides starts from the related L-amino acid.

Problems

1. Why did nature "waste" an ATP in glutamine biosynthesis? The lone pair of electrons on an ammonia could have attacked the γ-carbonyl group of a glutamate. Subsequent elimination of an oxide ion and release of a proton from the nitrogen would produce a glutamine without the consumption of an ATP.

2. Does it appear to be a paradox that L-glutamate is both the product and an initial reactant in the glutamine biosynthetic pathway? Assuming that glutamate synthase (fig. 21.3) is utilized, how can you explain this paradox?

3. Examine the bioenergetics of the synthesis of glutamine synthesis from α-ketoglutarate via glutamate synthase or glutamate dehydrogenase (fig. 21.3). Is there a difference?

4. What is the function of NADPH in the reactions catalyzed by glutamate dehydrogenase and glutamate synthase?

5. *E. coli* strains have been isolated that are unable to grow in a medium containing L-valine but lacking L-isoleucine and L-leucine. The same organism can grow on a medium lacking all three amino acids. Provide an explanation.

6. The amino acid 2-aminobutanoate is a product of some bacteria (not a protein component). Predict how the bacteria produces this amino acid.

7. A certain bacteria that was a tryptophan auxotroph was observed to grow well when it was supplied with tryptophan, but as soon as the tryptophan in the environment was exhausted it started to excrete a metabolite on the tryptophan biosynthetic pathway. Why didn't it excrete the metabolite before it exhausted the environmental tryptophan?

8. This chapter categorizes the amino acids into families based on the origin of their carbon skeleton. Is this an absolute pattern? Take a closer look at the information in this chapter to produce an answer.

9. What is the function of the acetylation of serine with acetyl-CoA during the biosynthesis of cysteine (fig. 21.8)?

10. Which ribose carbons are incorporated into tryptophan?

11. Molecules with structures as diverse as carbamoyl-phosphate, tryptophan, and cytidine triphosphate are feedback inhibitors of the *E. coli* glutamine synthase. The feedback inhibition is cumulative, with each metabolite exerting a partial inhibition on the enzyme. Why would complete inhibition of the glutamine synthase by a single metabolite be metabolically unsound?

12. Given the structural diversity of the compounds that feedback-inhibit glutamine synthase, would you predict that they interact at a common regulatory site? Why or why not?

13. How does increased synthesis of aspartate and glutamate affect the TCA cycle? How does the cell accommodate this effect?

14. In what sense may indole be viewed as an "intermediate" in L-tryptophan biosynthesis?

15. When ^{14}C-labeled 4-hydroxyproline was administered to rats, the 4-hydroxyproline in newly synthesized collagen was not radiolabeled. Explain.

16. The accumulation of biosynthetic intermediates, or of metabolites derived from these intermediates, has proven to be valuable in the analysis of biosynthetic pathways in microorganisms. It was found that these accumulations occurred only after the required amino acid had been consumed and growth had stopped. How might you account for this observation?

Solutions

1. The oxide ion is a very poor leaving group. Notice that the ATP actually contributes a phosphoryl group (PO_3 moiety) which is then eliminated as an HPO_4^{2-} upon the approach of an $:NH_3$. The function of the ATP is to assist in the removal of an oxygen from the carboxylate group.

$$\underset{\text{Glutamate}}{R-\overset{O}{\overset{\|}{C}}-O^-} + {}^-O-\underset{O^-}{\overset{O}{\overset{\|}{P}}}-O-\underset{O^-}{\overset{O}{\overset{\|}{P}}}-AMP \xrightarrow{ADP} R-\overset{O}{\overset{\|}{C}}-O-\underset{O^-}{\overset{O}{\overset{\|}{P}}}-O^- \xrightarrow[H_3\ddot{N}]{HPO_4^{2-}} \underset{\text{Glutamine}}{R-\overset{O}{\overset{\|}{C}}-NH_2}$$

3. Using glutamate synthetase the overall reaction is:

$$\alpha\text{-ketoglutarate} + \text{Gln} + 2\ \text{ATP} + \text{NADPH} + 2\ \text{NH}_3 \rightarrow 2\ \text{Gln} + 2\ \text{ADP} + 2\ \text{P}_i + \text{NADP}^+$$

or net:

$$\alpha\text{-ketoglutarate} + 2\ \text{ATP} + \text{NADPH} + 2\ \text{NH}_3 \rightarrow \text{Gln} + 2\ \text{ADP} + 2\ \text{P}_i + \text{NADP}^+$$

With glutamate dehydrogenase the overall reaction is:

$$\alpha\text{-ketoglutarate} + \text{ATP} + \text{NADPH} + 2\ \text{NH}_3 \rightarrow \text{Gln} + \text{ADP} + \text{P}_i + \text{NADP}^+$$

The two sequences differ by an ATP.

5. The availability of valine inhibited an enzyme in the early stages of valine biosynthesis (such as acetohydroxyacid synthetase, fig. 21.10). Because this inhibited enzyme is common to the biosynthetic pathways of isoleucine, valine, and leucine, the *E. Coli* strain did not grow because of a lack of leucine and isoleucine.

7. Why make more when you already have enough? The environmentally derived tryptophan inhibits the first enzyme in the pathway preventing flow of material through the pathway. Only when the environmental tryptophan is depleted will material flow through the pathway, or in this case because of the auxotroph only part way through the pathway.

9. In this two step pathway the acetyl group is added in the first reaction and an acetate leaves, while an HS^- is added in the second step. The function of the acetyl group is for the removal of the serine oxygen. The OH group is a poor leaving group; acetate is better. Also notice the O-acetyl homoserine in figure 21.8. Did nature have to use an acetyl group for the removal of oxygen? See the answer to question 1 of this chapter for an idea to answer this latter question.

11. Glutamine synthase catalyzes the formation of glutamine, the organically bound nitrogen of which is used in the biosynthesis of a number of structurally unrelated nitrogen-containing compounds. Glutamine synthase activity is well regulated, as you would predict for an enzyme at the beginning of a multibranched biosynthetic system. The *E. coli* glutamine synthase is regulated by feedback inhibition exerted by products of pathways dependent on glutamine, and by covalent modification.

    The metabolites carbamoylphosphate, tryptophan, and cytidine triphosphate, as well as glucosamine-6-phosphate, histidine, and AMPs, are synthesized by pathways that incorporate nitrogen directly from glutamine. The supply of glutamine must respond to the demand generated by the several pathways. Hence, total inhibition of the glutamine synthase by a single product could inhibit biosynthesis of critically needed end products from other pathways. Cumulative inhibition of the glutamine synthase by end products of the various pathways modulates the supply of glutamine in direct response to the demand.

13. Synthesis of aspartate by transamination of oxaloacetate and of glutamate by transamination of α-ketoglutarate depletes the concentration of oxaloacetate and α-ketoglutarate in the TCA cycle. Were these TCA cycle intermediates not replenished, the rate of acetyl-CoA oxidation and subsequent ATP production would be markedly diminished. Pyruvate, derived from the glycolytic metabolism of glucose, is used to replenish the concentration of each of the cycle intermediates. As noted in the previous solution, oxaloacetate is formed from the carboxylation of pyruvate, catalyzed by pyruvate carboxylase. Synthesis of citrate, from the condensation of oxaloacetate and acetyl-CoA, in turn replenishes the concentration of the other intermediates.

15. Hydroxylation of proline in collagen is a posttranslational modification. Proline, rather than 4-hydroxyproline, is incorporated into collagen precursors and is subsequently hydroxylated by proline-4 hydroxylase. Neither free proline nor proline bound to the prolyl-t-RNA are hydroxylated. The hydroxylated peptides are then assembled and processed into collagen.

22 Amino Acid Metabolism in Vertebrates

Summary

1. Only eight of the *de novo* pathways for amino acid biosynthesis can be found in humans. These amino acids are all related by a small number of steps to glycolytic or TCA cycle intermediates. A number of additional amino acids can be formed from these amino acids. Essential amino acids are those that must be supplied in the diet.
2. For most amino acids the α-amino group is removed at an early stage in catabolism, usually in the first step. Transaminases are specific for different amino acids. Frequently α-ketoglutarate is the acceptor for the amino group, in which case it is converted into glutamate. The α-ketoglutarate can be regenerated from the glutamate by oxidative deamination.
3. A great deal of excess NH_3 frequently results from amino acid catabolism. This excess ammonia must be eliminated. In bacteria and lower eukaryotes the ammonia can usually be removed by simple diffusion, but in higher eukaryotes this is not feasible. Since the ammonia is frequently quite toxic, it is detoxified before removal by conversion to urea or uric acid. An intricate pathway resulting in the conversion of ammonia into urea involves five enzymes. Three located in the cytoplasm and the remaining two in the mitochondrial matrix.
4. All amino acids can be degraded to CO_2 and water via the TCA cycle by the appropriate enzymes: The pathways often contain branchpoints to useful biosynthetic products. In every case, the pathways involve the formation of a dicarboxylic acid intermediate of the TCA cycle of pyruvate or of acetyl-CoA.
5. The discussion of amino acid catabolism is organized according to the common intermediates formed during degradation. Alanine, glycine, threonine, serine, and cysteine are degraded to acetyl-CoA by way of pyruvate. Threonine is also degraded to acetyl-CoA via pyruvate, but it also yields acetyl-CoA directly. Phenylalanine, tyrosine, tryptophan, lysine, and leucine also lead to acetyl-CoA, but they go by way of acetoacetyl-CoA rather than pyruvate. Arginine, histidine, proline, glutamic acid, and glutamine are all degraded to α-ketoglutarate. Catabolism of methionine, valine, and isoleucine leads to succinyl-CoA. Aspartate and asparagine are converted to oxaloacetate on degradation.
6. The importance of catabolic pathways is underscored by a broad spectrum of human metabolic diseases in each of which one enzyme for normal amino acid catabolism is either missing or defective.
7. Many biologically important routes of amino acid utilization, other than those leading to incorporation into proteins, are known. Some of these routes are distinctly anabolic pathways in which the amino acids serve as an initial substrate in an independent biosynthetic pathway. Other simple pathways involve the conversion of one amino acid to another, such as the formation of tyrosine from phenylalanine. The utilization of glycine in the formation of porphyrin derivatives occurs by very complex highly branched pathways. Some other biologically important pathways lead to the biosynthesis of small peptides as in the biosynthesis of glutathione.

Problems

1. The pathway for the biosynthesis of arginine (fig. 22.2) has several intermediates that have an *N*-acetyl group. Compare these intermediates with those in the salvage pathway (fig. 22.2) and propose a reason for the acetyl groups.

2. In the conversion of the *N*-acetyl-γ-glutamyl phosphate to *N*-acetylglutamic-γ-semialdehyde (fig. 22.2), two processes occur: An elimination of a phosphate and a reduction. Which step occurs first? Also can you propose a reason for the use of the phosphate group in the first place?

3. Often ammonia is portrayed as "feeding into" the urea cycle, for example, see figure 22.10. Do ammonia molecules "feed into" the urea cycle?

4. What effect does deprivation of dietary pyridoxal phosphate have on the capacity to metabolize amino acids?

5. The *de novo* biosynthetic pathway for the biosynthesis of arginine has been lost by higher animals. However the biosynthesis of arginine via a salvage pathway from proline can occur. What arguments can you make that it is unlikely that mammals would completely lose all capabilities to produce arginine?

6. Reduce the names "ketogenic" and "glucogenic" to the simplest possible terms.

7. Use figure 22.11 to deduce which amino acids are glucogenic, ketogenic, or both.

8. Many of the inborn errors in amino acid metabolism appear to result in mental retardation (table 22.3). The majority of these individuals appear normal at birth, but their mental capabilities fail to develop. Can you provide an explanation for this observation?

9. Why have many inborn errors been found in the metabolism of amino acids in humans but "none" in glycolysis, the citric acid cycle, or electron transport?

10. Fumarate is a product of both argininosuccinate lyase (fig. 22.7) and fumarylacetoacetase (fig. 22.12), even though the reactions are quite different. If you were describing these two reactions to an organic chemistry student using one word per reaction, what words would you pick?

11. The biosynthesis of δ-aminolevulinate is known to occur with the loss of one glycine carbon-bound hydrogen, producing the intermediate shown in figure 22.13. What coenzyme would you expect to participate in this process? How does the same coenzyme stabilize the carbanion formed in the decarboxylation part of the reaction?

12. What are the functions of the two water molecules that are utilized by the enzyme 5-oxoprolinase (fig. 22.15)?

13 Would you anticipate elevated arginase activity in the liver of an untreated diabetic animal? Why or why not?

14. The concentration of phenylalanine in the blood of neonates is used to screen for phenylketonuria (PKU). Explain the biochemical basis for the correlation of elevated blood phenylalanine concentration and PKU. Explain why restriction of dietary phenylalanine is critically important for youngsters with PKU.

15. a. L-Glutathione is not a primary gene product as are proteins. What "information" is used to direct the synthesis of L-glutathione?
    b. Predict the effect of a glutathione synthase inhibitor on cells exposed to oxidative stress.

Solutions

1. The N-acetyl groups in the *de novo* pathway prevent the spontaneous cyclization of glutamate-δ-semialdehyde. Acetylation provides a means of differentiating, for control purposes, similar chemical species involved in the de novo arginine pathway from those involved in the salvage pathway (fig. 22.2) and proline biosynthesis (fig. 21.1). Also, the intermediates are distinct for control purposes.
3. One of the nitrogens enters the urea cycle via carbamoyl phosphate which comes from ammonia (fig. 22.8). The second urea nitrogen enters the urea cycle as part of aspartic acid which in turn can come from transamination of oxaloacetate. Glutamate is the source of the amino group in the transamination. Glutamate in turn can arise from α-ketoglutarate by transamination or a reductive amination (glutamate dehydrogenase) involving ammonia.
5. Because of the importance of the urea cycle the capacity to convert ornithine into arginine is obvious. Complete loss of the ability to produce ornithine, (a catalyst or carrier in the urea cycle) would limit the organism's control over production of its nitrogen waste product.
7. Based on figure 22.11, Thr, Ala, Ser, Gly, Cys, Asn, Asp, Gln, Glu, His, Arg, Pro, Val, and Met are glucogenic; Lys, Trp, and Leu are ketogenic; and Phe, Tyr, and Ile are both ketogenic and glucogenic.
9. A homozygotic defect in a gene for a protein in glycolysis, the citric acid cycle, or electron transport would probably lead to the death of the cell(s) soon after fertilization because of their critical role in ATP production.

11. Pyridoxal phosphate forms a Schiff base (imine) with the glycine. A carbon bound hydrogen is labile and the resulting carbanion stabilized by resonance back into the pyridoxal phosphate. The carbanion approaches the carbonyl carbon of the succinyl-CoA. Following the elimination of the CoASH the intermediate shown in figure 22.13 is formed. The intermediate then loses a CO_2 forming a carbanion that is resonance stabilized back into the pyridoxal phosphate. Protonation of the latter intermediate produces δ-aminolevulinate.

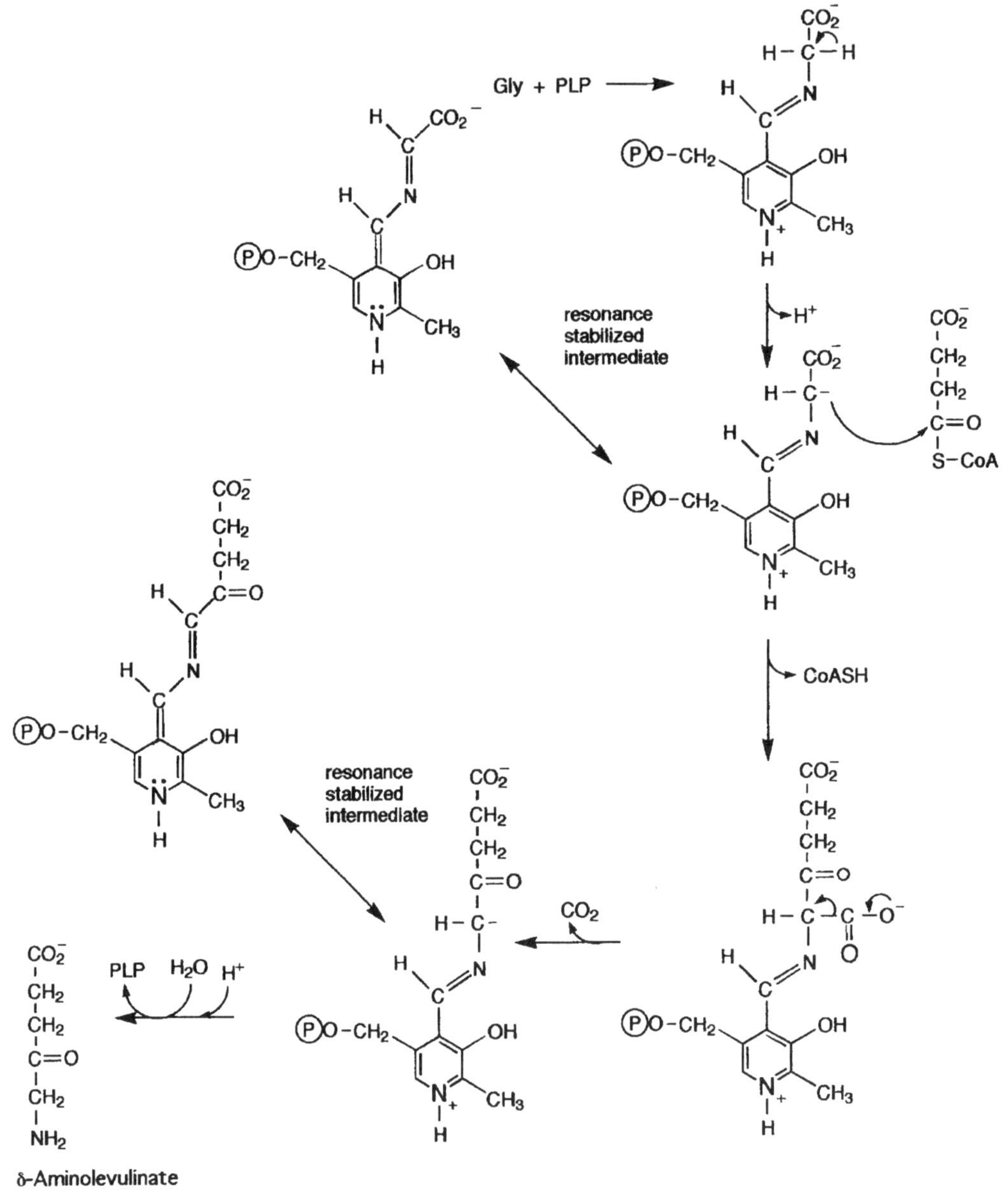

13. The untreated diabetic animal synthesizes glucose primarily by hepatic gluconeogenesis utilizing amino acids derived from protein catabolism as the carbon source. Recall that the ATP required for gluconeogenesis is generated by fatty acid β oxidation.

The amino acids are transaminated with α-ketoglutarate and oxaloacetate, forming glutamate and aspartate to be used in ureogenesis. The α-keto acids (those from glycogenic amino acids) then enter gluconeogenesis as pyruvate or as TCA cycle intermediates.

The urea cycle activity must increase to accommodate the increased flux of amino groups removed from the amino acids. Arginase catalyzes the hydrolysis of arginine yielding urea plus ornithine and is the rate-limiting step in the urea cycle. Hence you would predict an increased arginase activity in the liver of the untreated diabetic animal.

15. a. L-Glutathione is synthesized in successful steps catalyzed by γ-glutamylcysteine synthase and glutathione synthase. The information for the synthesis is dictated by the specificity of each enzyme. These enzymes are primary gene products (proteins), whereas L-glutathione is a secondary gene product. The formation of γ-glutamylglycine or cysteinylglycine is prevented because of the lack of the appropriate enzymes.

    b. Glutathione, a substrate for glutathione peroxidase, is used catalytically in the cell as an antioxidant but is also a substrate consumed stoichiometrically by glutathione transferases. Loss of glutathione by conjugation with certain xenobiotics slated for excretion could compromise the cell's antioxidant defenses. Hence the cell requires an adequate supply of glutathione. One role of glutathione as an antioxidant is shown

$$2\ GSH + H_2O_2 \rightarrow G\text{-}S\text{-}S\text{-}G + 2\ H_2O\ (Rxn\ 1)$$

    where GSH and G-S-S-G are reduced and oxidized glutathione, respectively.

$$G\text{-}S\text{-}S\text{-}G + NADPH + H^+ \rightarrow 2\ GSH + NADP^+\ (Rxn\ 2)$$

    Reaction 1 is catalyzed by glutathione peroxidase and reaction 2 by glutathione reductase. Thus, an inhibition of glutathione synthesis would lead to depletion of cellular GSH and an increased likelihood of oxidative damage, particularly if the cell was oxidatively stressed.

23 Nucleotides

Summary

Nucleotides are the building blocks for nucleic acids; they are also involved in a wide variety of metabolic processes. They serve as the carriers of high-energy phosphate and as the precursors of several coenzymes and regulatory small molecules. Nucleotides can be synthesized *de novo* from small-molecule precursors or, through salvage pathways, from the partial breakdown products of nucleic acids. The highlights of our discussion in this chapter are as follows.

1. The ribose for nucleotide synthesis comes from glucose, either by means of the pentose phosphate pathway or from glycolytic intermediates through transketolase-transaldolase reactions. Ribose-5-phosphate is converted to phosphoribosylpyrophosphate (PRPP), the starting point for purine synthesis. This pathway also incorporates into purines atoms from glycine, aspartate, glutamate, CO_2, and one-carbon fragments carried by folates. IMP synthesized by this route is converted by two-step pathways to AMP and GMP, respectively.
2. The biosynthetic pathway to UMP starts from carbamoyl phosphate and results in the synthesis of the pyrimidine orotate, to which ribose phosphate is subsequently attached. CTP is subsequently formed from UTP. Deoxyribonucleotides are formed by reduction of ribonucleotides (diphosphates in most cells). Thymidylate is formed from dUMP.
3. Several inhibitors of nucleotide biosynthesis are known. Each is extremely toxic, especially to rapidly growing cells, where the need for nucleic acid synthesis is greatest. In limited amounts, some of the inhibitors have chemotherapeutic value in the treatment of cancer and other illnesses. Some analogs of normal nucleosides are proving to be useful in the treatment of AIDS and certain other viral infections.
4. Nucleic acids and nucleotides are degraded to nucleosides or free bases before they are ingested. Nucleotides or their partial degradation products may be reutilized for nucleic acid synthesis, or they may be further catabolized for excretion or for use in the synthesis of other products. Purine nucleotides are degraded via guanine, hypoxanthine, and xanthine to uric acid, which in some species is degraded further before excretion. Inherited deficiencies in some of the enzymes involved in nucleotide degradation and salvage cause severe impairment of health, a fact testifying to the importance of the degradative pathways. Nucleotides of uracil and cytosine are degraded via uridine and uracil to simpler substances such as β-alanine.
5. All biosynthetic pathways are under regulatory control by key allosteric enzymes that are influenced by the end products of the pathways. For example, the first step in the pathway for purine biosynthesis is inhibited in a concerted fashion by nucleotides of either adenine or guanine. In addition, the nucleoside monophosphate of each of these bases inhibits its own formation from inosine monophosphate (IMP). On the other hand, adenine nucleotides stimulate the conversion of IMP into GMP, and GTP is needed for AMP formation.

Problems

1. How do you explain the observation that pyrimidine biosynthesis in bacteria is regulated at the level of aspartate carbamoyltransferase, whereas most of the regulation in humans is at the level of carbamoyl phosphate synthase?

2. Compare and contrast the pathways for the biosynthesis of IMP (fig. 23.10) with that of UMP (fig. 23.13). What are the two most striking features of this comparison?

3. The first enzyme in the biosynthesis of IMP (fig. 23.10) utilizes a glutamine to provide an amino group. We have seen the same phenomenon in the two previous chapters. Why doesn't nature use an ammonium ion rather than the more expensive (in the expenditures of ATP) glutamine as an amine donor?

4. Examine the reactants and products of the second enzyme on the IMP biosynthetic pathway. What reaction intermediate would you expect?

5. What function can you propose for the ATP in the fourth reaction in figure 23.10?

6. What function can you give ATP (in a mechanistic sense) in the fifth reaction (fig. 23.10) in the formation of IMP?

7. Examine figure 23.15 and deduce the fate of the oxygen on carbon two of the ribose residue.

8. What happens if you give allopurinol to a chicken?

9. Explain why Lesch-Nyhan patients suffer from severe gout. Although these patients can be treated with allopurinol to relieve the symptoms of gout, this treatment has no effect on the severe mental retardation. Suggest a possible explanation.

10. Why can the toxic effect of sulfanilamide on bacteria be reversed by *p*-aminobenzoate? Why are sulfa drugs not very toxic to humans?

11. Substitute thymine for uracil in figure 23.23 and determine the metabolites that would follow in the pathway.

12. Explain how antifolates like methotrexate selectively kill cancer cells. Why do cancer patients lose their hair, intestinal mucosa, cells of the immune system, and so forth when treated with antifolates?

Solutions

1. Carbamoyl phosphate synthetase contributes to two processes: a) the initial enzyme in the biosynthesis of pyrimidines and b) a component in the synthesis of arginine biosynthesis (fig. 22.1) or the urea cycle (fig. 22.7). In bacteria, both of these processes occur within the same compartment. If carbamoyl phosphate was controlled by the product(s) of the pyrimidine biosynthetic pathway then control of arginine biosynthesis would be lost. Therefore, in bacteria the second enzyme in the pathway (aspartate carbamoyltransferase) is controlled. In humans, the carbamoyl phosphate synthetase involved in the urea cycle is contained in the mitochondria (fig. 22.7), isolated from the cytosol counter part that is involved in the biosynthesis of pyrimidines. Because the two carbamoyl phosphate synthetases are in separate cellular compartments in humans there is no impact on the urea cycle by control of the cytosol carbamoyl phosphate synthetase by pyrimidine pathway products.
3. Typically, the mechanism for amine addition involves a nucleophilic approach by nitrogen. A lone pair of electrons is required on the nitrogen. Ammonium ions are protonated at physiological pH and do not have a lone pair. The amide of glutamine is not protonated and carries a lone pair of electrons.
5. At some point during the mechanism, a phosphoryl group is transferred from the ATP to the carbonyl carbon on the glycyl moiety of the 5′-phosphoribosyl-N-formylglycinamide. Later during the mechanism P_i is eliminated, taking with it what was the carbonyl oxygen. The phosphorylation converts the oxygen into a much better leaving group.
7. The conversion of the thioredoxin $(SH)_2$ into thioredoxin (S-S) produces two electrons and two hydrogen ions which ultimately combine with the 2′-oxygen on the ribose moiety to produce water.
9. Patients with Lesch-Nyhan syndrome have a deficiency in hypoxanthine-guanine phosphoribosyl transferase, an important enzyme in the salvage pathway for purines. When this enzyme is missing, the levels of phosphoribosylpyrophosphate (PRPP) become elevated and stimulate the synthesis of purines. These excess purines are degraded to uric acid, leading to severe gout. The gout can be treated with drugs like allopurinol, but this treatment has no effect on the severe mental retardation seen in these patients. The brain may lack a *de novo* purine biosynthetic pathway; it depends on the salvage pathway to produce purines, which are necessary for DNA replication. When a human is born, the rapid brain growth and development that normally occurs is inhibited if adequate levels of purine nucleotides are not produced. This genetic defect may eventually be treated with gene therapy.
11. The product would be 3-amino-2-methylpropanoic acid.

24 Integration of Metabolism and Hormone Action

Summary

In this chapter we focused on the ways in which various metabolic activities are integrated, with special attention to the nature and functioning of hormones. The following points are the highlights of our discussion.

1. Tissues store biochemically useful energy in three major forms: Carbohydrates, lipids, and proteins. Each tissue makes characteristic demands on and contributions to the energy supply of the organism.
2. Hormones are chemical messengers formed in specific tissues. They circulate between tissues of multicellular organisms and serve to coordinate metabolic activities, maintain homeostasis of essential nutrients, and prepare the organism for reproduction.
3. Most hormones fall into three classes: Polypeptides, steroids, and amino acid derivatives. Polypeptide hormones are synthesized from large precursors. Steroid hormones are derivatives of cholesterol. Thyroid hormones and epinephrine are amino acid derivatives.
4. A number of factors—synthesis, rate of release, and rate of elimination—determine the concentration of circulating hormone.
5. Hormones act by reversibly binding to proteins called receptors, an event that results in a conformational change that is detected by other macromolecules (acceptors) and that eventually leads to activation of rate-limiting enzymes. Each class of hormones binds to specific receptors that activate other membrane proteins.
6. Most membrane receptors generate a diffusible intracellular signal called a second messenger. Five intracellular messengers are currently known: Cyclic AMP, cyclic GMP, inositol triphosphate, diacylglycerol, and calcium. Second messengers usually activate or inhibit the action of one or more enzymes.
7. Steroid hormones penetrate the cell and bind to receptors in the nucleus, and activate (or sometimes repress) transcription of specific genes. Thyroid hormones act similarly.
8. A large number of diseases are due to either overproduction or underproduction of hormones, or to insensitivity of target tissues to circulating hormones. Knowledge of hormone biosynthesis, secretion, and interaction with target cells is essential to an understanding of the biochemical basis of these disorders.
9. In addition to classical hormones, other chemical messengers called growth factors serve to coordinate growth of tissues during development.
10. Plants make several hormones that regulate growth and differentiation.

Problems

1. Most metabolic conversions can occur in either direction by using different pathways. Eukaryotic cells frequently take advantage of subcellular compartments to separate oppositely directed pathways. Use fatty acid synthesis and degradation as an example, and discuss the design of the paths from the point of view that they are thermodynamically favorable and kinetically regulated (be sure to consider subcellular compartments). Discuss the design of glycolysis and gluconeogenesis in the liver.

2. If a starving person (one who has gone a number of weeks with no food) is given a shot of insulin, what happens? Explain your answer.

3. Why are all known hormone receptors proteins? Can other macromolecules serve as receptors?

4. What is the advantage of the fact that muscle tissue can use an anaerobic metabolism?

5. List several reasons why polypeptide hormones are synthesized as precursors.

6. A patient has a hypothyroid condition. He has low serum T_3 and T_4 levels and elevated levels of serum thyroid-stimulating hormone (TSH). After injection of thyrotropin-releasing hormone (TRH), his serum TSH goes even higher. Is his defect primary (thyroid), secondary (pituitary), or tertiary (hypothalamus)?

7. Thyrotropin-releasing hormone (TRH) contains an unusual amino acid, pyroglutamate (fig. 24.8). Have we encountered this amino acid before under a different name? Where do you think the name pyroglutamate came from?

8. Activation of most membrane-associated hormone receptors generates a second messenger. What is a second messenger and what are the five second messengers currently known? What role do G proteins play in second-messenger formation?

9. Is vitamin D a hormone or a vitamin? Explain your answer.

10. Inhibitors of protein synthesis have been shown to block both the rapid and slow auxin-mediated growth responses. How do you explain these observations?

Solutions

1. Pathways are designed so that the equilibrium is very favorable for the conversion. The same pathway cannot be used for conversion in either direction because if it were thermodynamically favorable in one direction, it would be unfavorable by the same amount in the opposite direction. Also, the simultaneous occurrence of conversions in both directions would serve no purpose, as it would result in a futile cycle and waste energy. Regulatory enzymes are designed so that pathways never operate simultaneously in both directions.

   Fatty acid synthesis and degradation demonstrate these principles. In a liver cell, fatty acid synthesis takes place in the cytosol using acetyl-CoA carboxylase and a large, multifunctional polypeptide fatty acid synthase. The first committed step is acetyl-CoA carboxylase, which is highly regulated by hormonal control; this results in the phosphorylation of the enzyme. Citrate is transported from the mitochondria and is used to generate acetyl-CoA and reducing power in the form of NADPH (NADPH is used in biosynthetic reactions instead of NADH). Citrate activates the carboxylase while the end product, palmitate, inhibits the reaction. Fatty acid degradation occurs in the matrix of the mitochondria. A key point of regulation is on the uptake of the fatty acid into the matrix of the mitochondria. Malonyl-CoA, the product of the acetyl-CoA carboxylase, inhibits uptake and prevents the newly made palmitic acid from being degraded.

Gluconeogenesis utilizes many of the glycolytic enzymes, yet three of these enzymes in glycolysis have large negative free energy changes in the direction of pyruvate formation. These reactions must be replaced in gluconeogenesis to make glucose formation thermodynamically favorable. This allows both glycolysis and gluconeogenesis to be thermodynamically favorable and at the same time permits the pathways to be independently regulated so that a futile cycle does not exist. The first step in gluconeogenesis involves the movement of pyruvate into the matrix of the mitochondria, where it is converted to phosphoenolpyruvate (PEP). PEP is transported back into the cytosol where it is converted by the glycolytic enzymes back to fructose-1,6-bisphosphate. Then two additional enzymes unique to gluconeogenesis allow glucose to be made. One of the most important allosteric effectors is fructose-2,6-bisphosphate, which activates phosphofructokinase, stimulating glycolysis, while at the same time inhibiting fructose bisphosphatase, inhibiting gluconeogenesis. The levels of fructose-2,6-bisphosphate are under hormonal regulation in response to blood glucose levels.

3. A hormone receptor must do two things if it is to function properly:
   a. Distinguish the hormone from all other surrounding chemical signals and bind it with a very high affinity (K_d ranges from 1×10^{-7} M to 1×10^{-12} M).
   b. Upon binding the hormone, undergo a conformational change into an active form that can then interact with other molecules that initiate the molecular events leading to the hormone's elicited response.

   Proteins are the only macromolecules that can exhibit this kind of behavior (specific binding and conformation change).

   Acceptors are macromolecules that react to a receptor's conformational change by mediating enzyme activation (or inactivation). This is done by specific phosphorylation of proteins or production of regulatory molecules, both enzymatic events. Therefore, acceptors must be proteins also. The only exception to this (in a loose sense) is a DNA sequence to which a hormone-receptor complex binds and whose perturbation affects a distant promoter and thereby alters the transcriptional activity of that gene.
5. The precursor-product relationship serves the cause of hormone function is a variety of contexts, some of which are listed here.
   a. A polypeptide signal sequence must be present if the protein is to be transported into the endoplasmic reticulum and subsequently secreted.
   b. Additional polypeptide sequences are necessary for proper peptide chain folding (e.g., C peptide of insulin).
   c. Cleavage allows control of hormones from inactive to active form (e.g., thyroxine).
   d. Production of a number of different hormones from the same precursor allows coordinate production of several hormones. Specific cleavage by the cell allows control of which peptides are produced (e.g., cleavage of prepro-opiocortin to corticotropin, β-lipotropin, γ-lipotropin, α-MSH, β-MSH, γ-MSH, endorphin, and enkephalin).
   e. A large precursor of the hormone can serve as a storage form (e.g., thyroglobulin).
7. The same amino acid, named as 5-oxoproline, is an intermediate in the γ-glutamyl cycle (fig. 22.15). The name pyroglutamate suggests formation involving dehydration via heat from glutamate.
9. Vitamin D can be considered both a hormone *and* a vitamin. Its mode of action is like many other steroid hormones (forming a receptor–hormone complex that activates transcription of specific genes in the nucleus), and it is synthesized in the body where it acts at a distant location. Vitamin D_3 is formed in the skin of animals through the action of ultraviolet light on 7-dehydrocholesterol. Vitamin D can also be taken in the diet (commonly as a vitamin supplement in milk) and is then considered a vitamin.

Supplement 1: Principles of Physiology and Biochemistry: Neurotransmission

Summary

In this chapter we examined the way in which signals are propagated along neurons and transmitted to other nerve or muscle cells. The following points are central to our discussion.

1. Action potentials are waves of depolarization and repolarization of the plasma membrane. In a resting nerve cell, the electric potential gradient ($\Delta\psi$) across the plasma membrane is about –70 mV, inside negative. This potential difference is generated mainly by the unequal rates of diffusion of K^+ and Na^+ ions down concentration gradients maintained by the $Na^+ - K^+$ ATPase.
2. The plasma membranes of electrically excitable cells contain specific ion channels that can switch between closed (nonconductive) and open (conductive) states. This switching is controlled by changes in $\Delta\psi$. When a neuron receives an electric stimulus that depolarizes the membrane locally by about 20 mV, Na^+ channels in this region switch to the open state. Na^+ flows into the cell through these channels, down its electrochemical gradient. This flow causes $\Delta\psi$ to drop rapidly to zero and to overshoot briefly to a positive value. A wave of depolarization spreads along the membrane as the changing $\Delta\psi$ triggers additional Na^+ channels to open. The Na^+ channels close again spontaneously in about 1 ms, and channels that are specific for K^+ open. K^+ efflux then returns $\Delta\psi$ to a value somewhat more negative than its original value.
3. The voltage-gated Na^+ channel has one major subunit (α) and several smaller subunits. In the α subunit there are four homologous domains, each of which probably has six transmembrane α helices. One of the transmembrane helices from each domain has negatively charged residues that probably line the pore across the membrane; another helix has an abundance of positively charged residues. Changes in $\Delta\psi$ could cause these helices to move with respect to each other and thus to open or close the pore.
4. The voltage-gated K^+-channel protein consists of a single subunit with six putative transmembrane α helices that are homologous to the α helices in the individual domains of the Na^+ channel. The intact channel probably is composed of four copies of the protein. Mutations of amino acid residues in a region between two of the transmembrane helices alter the ion specificity of the channel. The amino-terminal region of the protein is critical for the return of the channel to a closed state. This could involve a "ball" domain that swings on a "chain" to plug the channel.
5. Synaptic transmission between excitable cells is mediated by ligand-gated ion channels. The presynaptic neuron releases a neurotransmitter that diffuses to the postsynaptic cell. Binding of the neurotransmitter to a receptor in the plasma membrane of the postsynaptic cell causes a pore for ions to open, which can result either in depolarization of the membrane (initiating an action potential) or in hyperpolarization (inhibiting the firing of the cell), depending on the ion specificity of the channel.
6. The acetylcholine receptor, which mediates transmission at neuromuscular junctions and at some synapses, has the subunit composition $\alpha_2\beta\gamma\delta$. The five subunits form a pentagonal structure with a transmembrane channel at the center. The channel has a wide aperture in the synaptic cleft but narrows tightly in the center of the phospholipid bilayer. The side chains of a ring of leucine residues in the central region could form the gate that opens or closes to control the flow of ions across the membrane. Binding of acetylcholine to the extracellular region of the protein causes conformational changes that must be relayed to the gate.

Supplement 2: Principles of Physiology and Biochemistry: Vision

Summary

In this supplement we described the biochemical mechanisms involved in vision. The main points of our discussion are as follows.

1. Light rays entering the eye of a vertebrate are refracted by the cornea and focused on the retina. Rod and cone cells in the retina contain the visual pigments that are responsible for the initial response to light.
2. The light-sensitive protein complex in the rods, rhodopsin, consists of 11-*cis*-retinal bound as a Schiff's base to a protein, opsin. Rhodopsin is an integral constituent of membranes that form a stack of disks at one end of the cell. Cone cells, which are responsible for the perception of color, contain similar complexes in infoldings of the plasma membrane.
3. When rhodopsin absorbs light, the retinal isomerizes to the all-*trans*-isomer. This structural change initiates a series of transformations of the pigment-protein complex that result in an interaction with another protein, transducin. The interaction with rhodopsin causes one of transducin's subunits to take up a molecule of GTP in exchange for bound GDP and to react with a phosphodiesterase that hydrolyzes cGMP. This reaction activates the phosphodiesterase. The resulting drop in cGMP concentration leads to a decrease in the Na^+ permeability of the plasma membrane. A hyperpolarization of the membrane then triggers the transmission of a signal at the rod's synaptic junction to an adjacent neural cell.

25 Structures of DNA and Nucleoproteins

Summary

1. The genetic material of cells and viruses consists of DNA or RNA. That DNA bears genetic information was first shown when the heritable transfer of various traits from one bacterial strain to another was found to be mediated by purified DNA.
2. All nucleic acids consist of covalently linked nucleotides. Each nucleotide has three characteristic components: (1) a purine or pyrimidine base; (2) a pentose; and (3) a phosphate group. The purine or pyrimidine bases are linked to the C-1′ carbon of a deoxyribose sugar in DNA or a ribose sugar in RNA. The phosphate groups are linked to the sugar at the C-5′ and C-3′ positions. The purine bases in both DNA and RNA are always adenine (A) and guanine (G). The pyrimidine bases in DNA are thymine (T) and cytosine (C); in RNA they are uracil (U) and cytosine. The bases may be post-replicatively or posttranscriptionally modified by methylation or other reactions in certain circumstances.
3. DNA exists most typically as a double-stranded molecule, but in rare instances it exists (in some phages and viruses) in a single-stranded form. The continuity of the strands is maintained by repeating 3′, 5′-phosphodiester linkages formed between the sugar and the phosphate groups; they constitute the covalent backbone of the macromolecule. The side chains of the covalent backbone consist of the purine or pyrimidine bases. In double-stranded, or duplex, DNA the two chains are held together in an antiparallel arrangement. The base composition of DNA varies characteristically from one species to another in the range of 25%-75% guanine plus cytosine. Specific pairing occurs between bases on one strand and bases on the other strand. The complementary base pairs are either A and T, which can form two hydrogen bonds, or G and C, which can form three hydrogen bonds. The duplex is stabilized by the edge-to-edge hydrogen bonds formed between these planar base pairs and face-to-face interactions (stacking) between adjacent base pairs. Twisting of the duplex structure into a helix makes stacking interactions possible.
4. The right-handed helical structure, known as B DNA, is the most commonly occurring conformation of linear duplex DNA in nature. In this structure, the distance between stacked base pairs is 3.4 Å, with approximately 10 base pairs per helical turn. The inherent flexibility of the structure, however, makes a variety of conformations possible under different conditions. In some instances, nucleotide sequence and degree of hydration dictate which conformations are favored. DNA interacts with a variety of proteins inside the cell, and these proteins can also have a significant influence on its secondary and tertiary structure.
5. Circular DNA molecules, which are topologically confined so that their ends are not free to rotate, can form supercoils that are either right-handed (negative) or left-handed (positive). Negative supercoiling exerts a torsional tension favoring the untwisting of the primary right-handed double helix, whereas positive supercoiling has the opposite effect. Negatively supercoiled DNAs are most commonly observed in prokaryotes, which contain an enzyme that generates the supercoiled structure.
6. When duplex DNA (or RNA) is heated, it dissociates (denatures) into single strands. The temperature at which denaturation occurs (the melting temperature) is a measure of the stability of the duplex and is a function of the G-C content of the DNA. A preparation of denatured DNA may be renatured in the native duplex structure by maintaining the temperature about

25°C below the melting temperature. The rate of renaturation is a measure of the sequence complexity of the DNA. In prokaryotes, which consist predominantly of unique sequences, the complexity (the number of base pairs) is approximately equal to the genome size. However, eukaryotic cells contain DNAs of varying sequence complexity that renature at quite different rates. The fastest renaturing fractions are present in many copies per nucleus, whereas the slowest renaturing fractions are present in single copies. Analyses by other techniques have shown that some of the repetitive DNA sequences exist as tandemly repeated structures, while other types of repetitive sequences are dispersed throughout the genome.

7. In the chromatin of eukaryotic cells DNA forms a coiled-coil structure with an approximately equal weight of a mixture of five basic proteins known as histones. Four of these histones in pairs form an octamer around which the DNA duplex occurs in a left-handed helix. The DNA octamer complex is called a nucleosome. Each nucleosome contains about 140 base pairs of DNA in a nuclease-resistant "nucleosome core" and approximately 60 base pairs of spacer between core particles. Histone H1 binds to the chromatin independently of the octamer and is the first histone to dissociate from the chromatin when the ionic strength is raised. Beyond the nucleosome the higher order structure of the chromosome involves coiled-coil structures with varying degrees of regularity.

Problems

1. Briefly describe how Avery was able to show that DNA is the genetic material in cells.

2. Why is the bond holding nucleotides together in nucleic acids called a phosphodiester bond?

3. Give the structure of the DNA strand complementary to pTpApCpG (see structure shown in fig. 25.3) in the abbreviated form.

4. How many different base-paired structures with two hydrogen bonds can be made using guanine (G) and thymine (T)? Which one of these is most similar to the standard Watson-Crick (G-C) base pair?

5. a. List the hydrogen bond donors and acceptors available in the major and minor grooves of the DNA double helix.
   b. The intermolecular forces that stabilize DNA-protein complexes often involve hydrogen bonds between specific amino acids and the exposed surfaces of the bases. Explain why interaction with the major groove is more common among DNA-binding proteins than interaction with the minor groove.

6. How many base pairs are found in the DNA pictured at the right in figure 25.10 if two negative supercoils are introduced without breaking the backbone? Describe the effect of a short stretch of Z DNA on the overall conformation of this DNA.

7. Describe a physical method that can be used to estimate the base composition of DNA. Describe the data obtained with two DNA samples: One with high G-C content and another with high A-T content? (Assume that the concentration of the samples is equal.)

8. Why is DNA denatured at either low pH (pH 2) or high pH (pH 11), and why is DNA stable at pH 7? (*Hint:* See p*K* values in table 23.2.)

9. You are given a sample of nucleic acid extracted from a virus. How would you determine whether the virus has an RNA or DNA genome and whether it is single- or double-stranded?

10. A column packed with the material hydroxyapatite preferentially binds double-stranded DNA over single-stranded DNA. The double-stranded DNA can be eluted by changing the salt concentration. Using a hydroxyapatite column and the information in figure 25.15, propose a method for separating a mixture of equal amounts of T4 bacteriophage and *E. coli* DNAs into the respective components.

11. Why can't RNA duplexes or RNA-DNA hybrids adopt the B conformation?

12. Some DNA-binding proteins specifically bind to Z DNA. How could these proteins help stabilize DNA in the Z configuration? (*Hint:* How do single-stranded DNA-binding proteins destabilize duplex structures?)

13. Linear duplex DNA can bind more ethidium bromide than covalently closed circular DNA of the same molecular weight. Why? (*Hint:* Ethidium bromide molecules bind between adjacent base pairs of DNA, causing the duplex to unwind in the region of binding.)

14. Give the relative times for 50% renaturation of the following pairs of denatured DNAs, starting with the same initial DNA concentrations.
    a. T4 DNA and *E. coli* DNA, each sheared to an average single-strand length of 400 nucleotides.
    b. Unsheared T4 DNA and sheared T4 DNA.

15. When histone proteins are isolated from chromatin their mass is equal to the DNA, and the molar ratio of four of the histones is 1:1:1:1 (H2a:H2b:H3:H4), while H1 is found in half the amount (0.5). Discuss whether or not these data fit the bead-and-string model for nucleosomes.

16. Your supervisor shows you two test tubes and announces: "The labels fell off of these tubes in the freezer! One was supposed to contain DNA from *E. coli* and one was DNA from *Mycobacterium tuberculosis.* I can't figure out which is which, so you will just have to grow cells and prepare more DNA." Because *M. tuberculosis* is a dangerous pathogen, you wish to avoid culturing the cells. Can you devise a simple strategy to determine which sample is from which organism, based on the fact that *E. coli* DNA contains 52% G + C and *M. tuberculosis* DNA is 70% G + C? Show an example of the expected results.

Solutions

1. Avery, working with two different strains of pneumococcus, was able to show that a fraction isolated from the pathogenic S strain that transformed the nonpathogenic R strain was DNA. This transforming activity was not affected by RNase, proteases, or enzymes that degrade capsular polysaccharides but was destroyed by treatment with DNase. Purified DNA from S cells was able to transform R cells into S cells *in vitro*.
3. By convention, nucleotide sequences are always given in the 5′ to 3′ direction. The structure shown in figure 25.3 is that of dpTpApCpGp. Because the two strands are antiparallel, the base at the 5′-end of the complementary strand would be the cytosine that forms a base pair with the guanosine at the 3′-end of the sequence given. The entire complementary strand would therefore be given as dpCpGpTpA, or, in more abbreviated form, as CGTA. Another shorthand version of this sequence is:

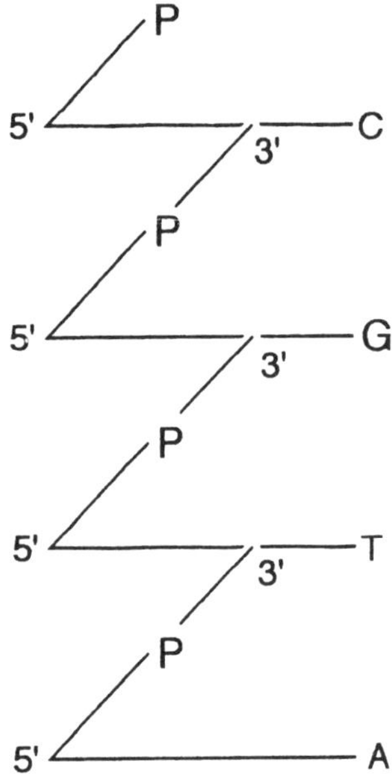

[The answer ATGC is incorrect, because the direction of the sequence is incorrect. 3′-ATGC-5′ is acceptable, because the correct direction is indicated.]

5. a. Note the designation of major and minor grooves in figure 25.6. The following hydrogen bond donors and acceptors are available in the major groove of B-form DNA:

Adenine:	N^7; N^6
Cytosine:	N^4
Guanine:	N^7, O^6
Thymine:	O^4

The following hydrogen bond donors and acceptors are available in the minor groove of B-form DNA:

Adenine:	N^3
Cytosine:	O^2
Guanine:	N^3; N^2
Thymine:	O^2

(Note also the possibility of van der Waals interactions in the major groove with C^5 and C^6 of the pyrimidines and C^8 of purines, and in the minor groove with C^2 of adenine.)

b. Although the numbers of electronegative atoms capable of forming hydrogen bonds in the major and minor groove are similar, access to those in the minor groove is hindered by the ribose moiety (see fig. 25.8). Because hydrogen bond donors and acceptors are more accessible in the major groove, most (but not all) proteins interact with this face of the DNA double helix.

7. The melting temperature (*T*m) of DNA is also affected by the base composition, with the G-C-rich DNA having a higher *T*m than the A-T-rich DNA. The DNA samples could be heated in a spectrophotometer and the increase in absorbance of ultraviolet light (hyperchromism) could be plotted against temperature. The A-T-rich DNA would have a lower *T*m than the G-C-rich DNA. Also, if the DNA melting was monitored at 260 nm, the A-T-rich DNA would show a larger hyperchromic increase in absorbance than an equal amount of G-C-rich DNA. If the DNA denaturation was followed at 280 nm, the G-C-rich DNA would have a larger hyperchromic increase.

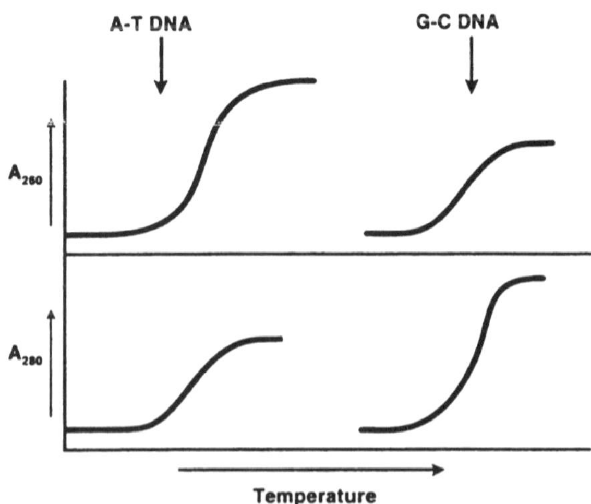

9. A simple approach would be to take small aliquots of the sample and treat with the enzymes ribonuclease (RNase) or deoxyribonuclease (DNase). Digestion of the sample by one of these enzymes would indicate whether the sample is RNA or DNA. Another method is to treat a small sample with alkali, which degrades RNA to mononucleotides, but only denatures DNA to the single-stranded form. One way to detect whether these treatments had any effect on the nucleic acid is to subject it to electrophoresis on an agarose gel. Free nucleotides, or even small oligonucleotides, are not visible on agarose gels, while the original viral nucleic acid should yield one (or more) discrete high molecular weight bands.

To determine whether the nucleic acid is single-stranded or double-stranded, it could be heated. A very sharp increase in absorbance at 260 nm would indicate a double-stranded RNA or DNA, while a broader melting curve would suggest a single-stranded nucleic acid. One could also analyze the nucleotide composition of the nucleic acid. Equivalence between A and T (or U) and between G and C would strongly suggest that the nucleic acid is double-stranded, while differing values of the nucleotides would suggest that the nucleic acid is single-stranded.

11. RNA duplexes and RNA-DNA hybrids adopt the so-called "A" conformation, a wider and flatter right-handed helix with tilted base pairs. As a result, the 2′-OH group found on the ribose in RNA sterically prevents the B duplex from forming.

13. The differences in Ethidium (Et) binding capacities between linear duplex DNA and covalently closed circular DNA (cccDNA) can be understood in terms of the differences in topological constraints imposed on the two molecules. By intercalating between two adjacent base pairs, Et unwinds the double helix, which results in an increase in the length of the helix (pitch). For cccDNA, the conformational stress introduced by unwinding is compensated for by a change in tertiary structure of the molecule, i.e., supercoiling. The winding and unwinding (twist) of a cccDNA molecule is related to supercoiling (writhe) in the following way:

$$\Delta L = \text{twist } (T) + \text{writhe } (W)$$

The linking number L remains constant provided that no covalent bonds are broken and re-formed, as in the present case. Since unwinding reduces T, W must assume more positive values in order to satisfy the relationship. Simply stated, unwinding the helix results in the introduction of positive supercoils.

At some point, the torsional stress caused by the positive supercoils will become energetically unfavorable and the tendency of the molecule will be toward winding, thus preventing further binding of Et. The tendency toward winding in positively supercoiled DNA is driven by the excess free energy of the supercoiled conformation resulting from torsional stress (as evidenced by the fact that introducing a nick into a supercoiled molecule completely relaxes the DNA). Linear duplex DNA and nicked circular DNA do not experience this torsional stress, since they are not covalently closed, and would thus be expected to have a greater binding capacity for Et.

It should be noted that if one starts with negatively supercoiled DNA (as opposed to *relaxed*, covalently closed circular DNA) and adds increasing concentrations of Et, the negatively supercoiled DNA will initially bind more Et than will duplex linear DNA. This is so because the negatively supercoiled conformation energetically favors unwinding of the helix and thus intercalation of Et. Ultimately, however, when Et is present in excess concentrations, linear duplex DNA has a greater capacity for Et binding as a result of the topological constraints imposed on cccDNA discussed earlier.

15. The ratio of the histones in chromatin supports the model proposed for nucleosome structure. The core of the nucleosome is made up of an octamer of two molecules each of H2A, H2B, H3, and H4. The H1 histone seals off the nucleosome, i.e., only one H1 per nucleosome. Thus, the ratio of histones fits the model proposed for nucleosome structure.

26 DNA Replication, Repair and Recombination

Summary

This chapter deals with reactions involved in DNA synthesis, degradation, repair, and recombination. The chief points to remember are as follows.

1. DNA replication proceeds by the synthesis of one new strand on each of the parental strands. This mode of replication is called semiconservative, and it appears to be universal. DNA synthesis initiates from a primer at a unique point on a prokaryotic template such as the *E. coli* chromosome. From the initiation point, DNA synthesis proceeds bidirectionally on the circular bacterial chromosome. The bidirectional mode of synthesis is not followed by all chromosomes. For some chromosomes, usually small in size, replication is unidirectional.
2. In eukaryotic systems, replication can start at several points (still not well defined) along the chromosome. Replication is usually bidirectional about each initiation site. The termination points of replication are interspersed between initiation sites. In most cases of unidirectional or bidirectional replication, synthesis occurs nearly (but not exactly) simultaneously on both strands of the parent DNA template. Because synthesis can occur only in the $5' \rightarrow 3'$ direction on the growing chain, and the two strands in the parent duplex are oriented in opposite directions, synthesis can occur continuously on only the leading strand. On the other (lagging) strand it must pause for the template to unwind. Synthesis on the lagging strand does not occur continuously but rather in small discontinuous spurts, generating Okazaki fragments.
3. Many proteins are required for DNA synthesis and chromosomal replication. These include polymerases; helicases, which unwind the parental duplex; enzymes that fill the gaps and join the ends in the case of lagging-strand synthesis; enzymes that synthesize RNA primers at various points along the DNA template; topoisomerases, which permit rotation and supercoiling; and single-strand DNA-binding proteins, which stabilize single-stranded regions that are transiently formed during replication. Most of these proteins have been isolated from whole cells and studied in cell-free systems.
4. In *E. coli*, mutations have been isolated in the genes encoding a number of these enzymes. Many of these mutations are conditional because the functional enzymes involved are required for DNA synthesis and cell viability. Mutants carrying mutationally altered proteins have been important in confirming their roles predicted from cell-free studies.
5. In eukaryotes, considerable progress has been made in studying the *in vitro* replication of animal viruses, such as SV40. The importance ascribed to the enzymes that have been characterized is largely based on a comparison of their properties with similar prokaryotic enzymes whose functions are better understood.
6. Many enzymes that act on DNA are involved in processes other than DNA synthesis. They include DNA repair enzymes, DNA degradation enzymes, and DNA recombination enzymes.
7. Enzymes that catalyze the synthesis of DNA using an RNA template are known as reverse transcriptases. The first reverse transcriptase discovered was encoded by an RNA retrovirus. This enzyme is needed in the virus replication cycle. Some animal viruses pass through an RNA intermediate and also require a reverse transcriptase to replicate the viral DNA. Similarly, a number of transposable elements found in cellular chromosomes replicate through RNA

intermediates; they usually encode a reverse transcriptase. A unique reverse transcriptase called telomerase is used to synthesize the DNA at the ends of linear eukaryotic chromosomes.

Problems

1. In the Meselson-Stahl experiment illustrated in figure 26.2, a sample of the DNA shown in tube 4 (labeled with ^{15}N followed by one generation in ^{14}N) was heat-denatured prior to being subjected to centrifugation in a CsCl density gradient. This gradient showed two peaks of single-stranded DNAs of different densities. How did this experiment further support the idea of semiconservative replication?

2. Explain why Cairns and coworkers used [^3H]-thymidine to label replicating *E. coli* DNA in the experiments shown in figure 26.3.

3. Draw the chemical reaction mechanism for the formation of a phosphodiester linkage during DNA synthesis. Discuss the significance of the pyrophosphate product that is formed. What is the significance of the Mg^{2+} requirement?

4. Cairns and De Lucia isolated a mutant strain of *E. coli* that had only about 1% of the DNA polymerase activity found in wild-type cells, yet the strain replicated its DNA at a normal rate. Explain how this discovery was important in understanding the role of the different DNA polymerases in replication and repair.

5. All enzymes that make DNA in a template-dependent fashion require a primer. How does the use of a primer increase the fidelity of DNA synthesis, and why is this primer usually RNA?

6. Even with its proof-reading activity, *E. coli* DNA polymerase III still exhibits a measurable rate of nucleotide misincorporation (about one mistake per 10^{10} nucleotides incorporated). Mutants of *E. coli* DNA polymerase III can be isolated that have a lower than normal rate of misincorporation. Why might such mutants, which can be said to have hyperaccurate DNA replication, be evolutionarily unfavorable?

7. *E. coli* has a genomic complexity of about 4×10^6 bp, and each replication fork can move at a rate of about 10^3 bp/s (base pairs per second). How long does it take to replicate the *E. coli* chromosome? With an ample carbon source and ideal growth conditions, cells of *E. coli* can divide in about 20 min. How can this shorter division time occur if the rate of fork migration remains constant at 10^3 bp/s?

8. Draw the chemical reaction mechanism for DNA ligase in *E. coli* (uses NAD^+ as the source of energy and forms a covalent intermediate with an ε-amino group of lysine). Why are ligation reactions that require ATP more thermodynamically favorable?

9. Humans have about 2.9×10^9 bp of DNA in their genome, and the replication fork migrates at the rate of approximately 30 bp/s. How long would it take to replicate the entire genome if it was a single continuous piece of DNA? How many replication origins are required to replicate this DNA in one hour?

10. Why does the reaction catalyzed by the *dnaB* gene product require ATP hydrolysis? Does this protein alter the linking number of DNA? Explain.

11. Eukaryotic DNA is replicated at a slower rate than prokaryotic DNA. One reason may be the requirement for the deposition of histone proteins on DNA (histone synthesis and DNA replication are coupled). Describe a model for the replication of eukaryotic DNA and nucleosome formation.

12. The graph shows *E. coli* labeled with radioactive thymidine for a short pulse (10 s) followed by a chase with an excess of nonradioactive thymidine. The DNA is then extracted and centrifuged in alkaline sucrose gradients (under high pH conditions the DNA denatures). Explain what these data imply, and interpret these results in light of our current model for DNA replication.

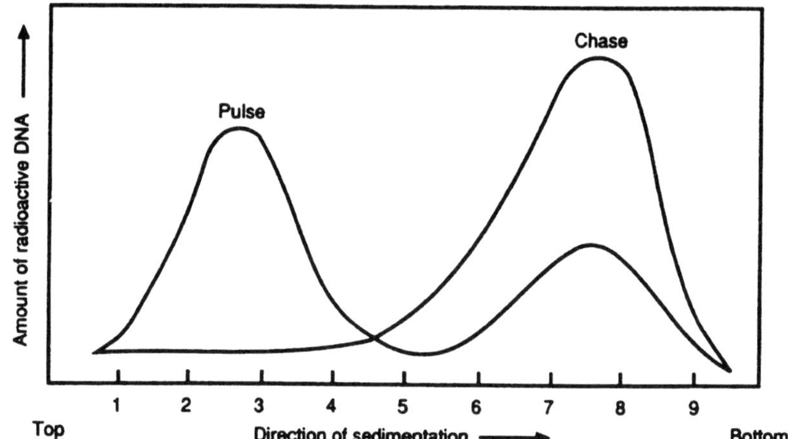

13. How is the SOS response reversed following repair of DNA?

14. Compare and contrast the excision repair and photoreactivation mechanisms for correction of ultraviolet-induced thymine dimers.

15. What is the role of the recA protein in *E. coli* DNA repair and in recombination? What use are *recA* mutants in biochemical and genetic research?

16. Normal human fibroblasts are grown in culture and then exposed to UV light. A short time later the DNA is extracted and applied to an alkaline sucrose gradient, and the data in graph (*a*) are observed. Another sample of cells is also exposed to UV light, but about 12 h are allowed to pass before the DNA is extracted and applied to an alkaline sucrose gradient (graph *b*). Explain these data from what you know about DNA repair (*Hint:* see fig. 26.18). Another sample of fibroblast cells, isolated from a patient with xeroderma pigmentosum (a disease resulting from the inability to repair DNA damage caused by UV light) was exposed to UV light and then applied to a gradient after a short time. Would you expect the data to resemble those in graph (*a*) or (*b*)? Why?

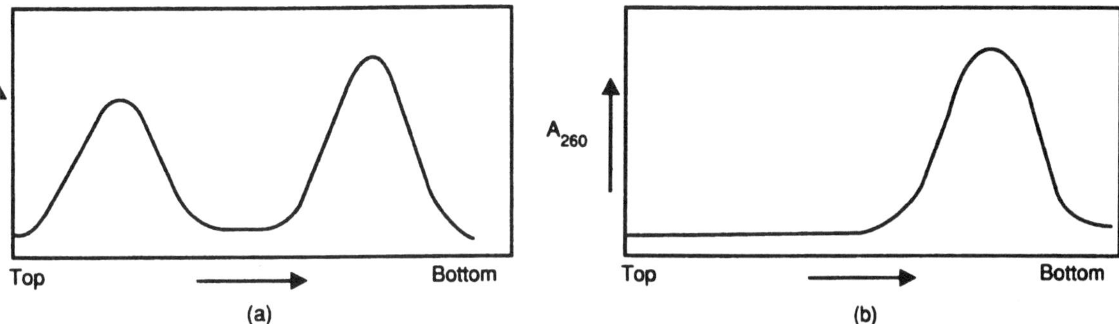

Solutions

1. If DNA replication were dispersive, then each strand of each daughter molecule would have had an intermediate density in this control experiment. The heat-denatured DNA would show only one peak in the CsCl gradient, of the same density as the sample in tube #4. The fact that after one generation, one strand of the DNA still had the same density as the parental (^{15}N) DNA served to further corroborate the conclusion that DNA replication is semiconservative.
3. The mechanism for the formation of a phosphodiester linkage during DNA synthesis is shown:

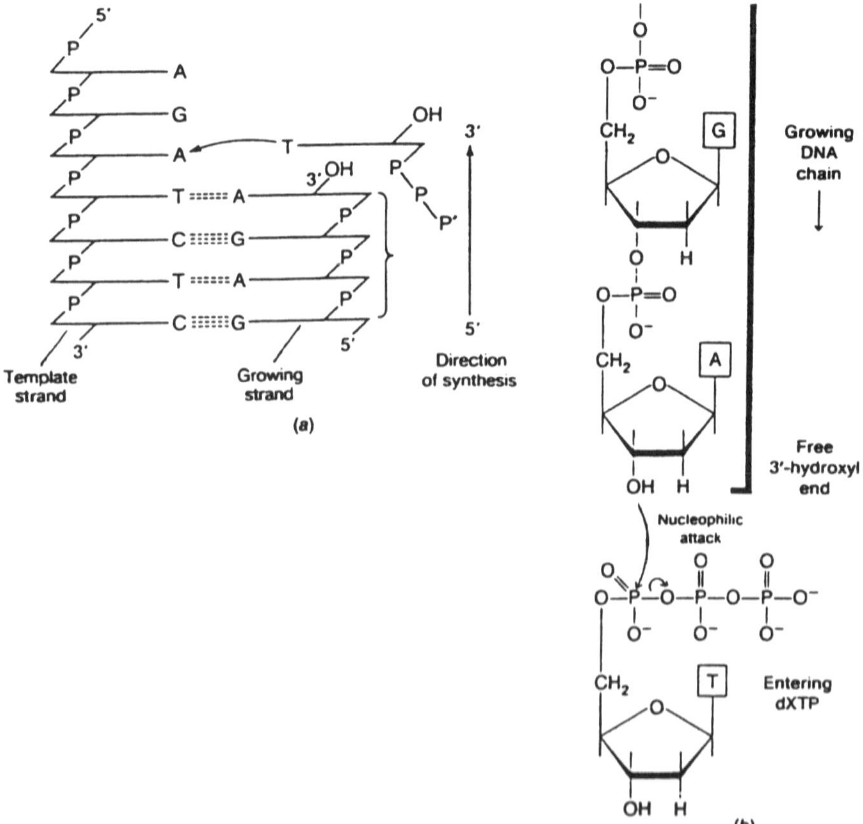

As has been seen in many other biochemical reactions in the cell, the generation of pyrophosphate coupled with its hydrolysis by pyrophosphatase is the major driving force for DNA synthesis. The Mg^{2+} can bind to the transition state intermediate (trigonal bipyramidal intermediate) and stabilize it, lowering the energy of activation. The metal ion can also promote the reaction by charge shielding; the Mg^{2+}NTP complex is the actual substrate, with the metal reducing the negative charge on the phosphate groups so as not to repel the electron pair of the attacking nucleophile.

5. Short chains of DNA are inherently more difficult to synthesize accurately than long chains. Hydrogen bonds are weak, and because of the lack of long-range stacking interactions, a few base-pairs of DNA do not form a very stable structure even when the bases are correctly paired. (Another way of saying this is that base-pairing is a cooperative process.) As a result, it would be difficult for a polymerase to proof-read those first few bases, if they were synthesized in the absence of a primer. The use of a primer increases fidelity in DNA synthesis by providing a more extensive stacked double helix to which the first few bases of DNA are added, allowing the proof-reading exonuclease activity of the polymerase to more accurately evaluate the stability of the newly synthesized DNA.

   Since the primer is not synthesized by a proof-reading polymerase, it is likely to contain some errors. It is apparently more efficient to remove the entire primer and synthesize new DNA across the resulting gap than to carefully repair the primer-containing region. The reason why the primer is made of RNA and not DNA may be because RNA, even in a double-stranded nucleic acid, is readily recognized as being different from DNA (RNA-DNA duplexes adopt a different conformation due to the presence of the 2′-OH on the ribonucleotides.) This recognition facilitates degradation of the RNA primer and its replacement by a proof-reading DNA polymerase.

   It is also thought that, very early in evolution, RNA was used as the genetic material, and not DNA. The enzymes for making RNA primers might have been present already when the very earliest DNA polymerases evolved and were natural candidates for primer synthesis.

7. It would take *E. coli* about 33 min to replicate its genome ($4 \times 10^6 / 2 \times 10^3 = 2 \times 10^3$ s or about 33 min). Since a replication fork moves at 10^3 bp per second and because there are two replication forks, the DNA is being replicated at 2×10^3 bp per second. Cells would divide faster if one round of replication started before the other finished, generating multiple replication forks (multifork replication) with division times of about 20 min.

9. 2.9×10^9 bp × 1 sec/60 bp × 1 h/3600 sec = 13,400 hours. 60 bp/sec is used in this calculation, because it is assumed that one origin of replication would generate two replication forks moving in opposite directions, each at a rate of 30 bp per sec. The human genome would need at least 13,400 replication origins to be completely replicated in one hour.

11. DNA in eukaryotic chromosomes is complexed with histone proteins in complexes called nucleosomes. These DNA-protein complexes are disassembled directly in front of the replication fork. The nucleosome disassembly may be rate-limiting the migration of the replication forks, as the rate of migration is slower in eukaryotes than prokaryotes. The length of Okazaki fragments is also similar to the size of the DNA between nucleosomes (about 200 bp). One model that would allow the synthesis of new eukaryotic DNA and nucleosome formation would be the disassembly of the histones in front of the replication fork and then the reassembly of the histones on the two duplex strands. Histone synthesis is closely coupled to DNA replication.

13. In the SOS response, recA protein, upon binding of damaged DNA fragments, acts as a protease and cleaves lexA repressor, releasing all of the genes normally repressed by lexA from its control.

Among the genes that are normally repressed by lexA protein are its own gene and that for recA protein (see fig. 26.19). Thus, when the SOS response is activated, the synthesis of lexA and recA proteins increases. As long as newly synthesized recA recognizes and binds damaged DNA, it will continue to cleave all of the lexA as it is made, and lexA will not act as a repressor. Even though the rate of synthesis of lexA is high, it is so rapidly cleaved by recA that there is no significant accumulation of intact lexA. The SOS response is reversed when the protease activity of recA can no longer be activated because most or all of the damaged DNA has been repaired or eliminated, and intact lexA protein levels begin to rise. Then lexA can act as a repressor, binding to the gene control regions of all of the genes it regulates, including its own and that of recA.

15. The recA protein is very important in DNA repair. An insult to the DNA leads to the activation of the protease function of recA, which then cleaves lexA protein, turning on the genes in the SOS response. Once the DNA is repaired, the recA protease is inactivated and new lexA protein is made, repressing the DNA repair genes again. RecA mutations are totally deficient in homologous recombination, demonstrating the important role of the recA protein in this process. The purified recA protein will catalyze the exchange between duplex and single-stranded DNAs with the hydrolysis of ATP. Also, recA protein will form a complex between two circular helices if one helix is gapped on one strand. These functions of rec A protein would place it at the hub of activities in recombination. RecA mutants would be very useful in genetic research because mutants generated would not be repaired, and in recombinant DNA cloning recombinational events between vector recombinant DNA and host genomic DNA would not occur.

27 DNA Manipulation and Its Applications

Summary

1. Sequencing DNA uses chemical methods to cleave specific bases in preexisting DNA or carries out the synthesis under conditions where synthesis is interrupted at specific bases.
2. A specific segment of DNA can be synthesized *in vitro* by the polymerase chain reaction. Short segments of DNA bordering the segment of interest are added to a mixture containing the segment of interest, a DNA polymerase, and the deoxyribotriphosphate substrates. The DNAs are first denatured, then annealed, and then synthesized. This cycle is repeated 20 or more times by raising the temperature to stop synthesis and lowering the temperature for annealing and synthesis. The outcome is a mixture in which the vast majority of the DNA is newly synthesized DNA bounded by the sequences of the added primers.
3. Another procedure for amplification cuts DNA containing the segment of interest into small pieces with a restriction enzyme. The cut pieces are incorporated into a plasmid or virus "vector" to be amplified in a suitable host. After growth, the mixture is plated to produce a mixture of bacterial or viral clones. The clone or clones of interest are identified often by hybridization of the clones after replica plating with a radioactive probe, followed by autoradiography to find the clone of interest.
4. Most cloning has been done in *E. coli*. Yeast is the most used eukaryotic host. Cloning is also possible in a number of plant and animal cells.
5. Mapping with recombinant DNA probes was first applied to the human globin genes. Starting probes were obtained by isolating the globin messenger from reticulocytes and converting it into a cDNA, which was used to scan a human genomic library for cross hybridizing members. Once detected and purified, those cross hybridizing members carrying globin messenger sequences were themselves converted to radioactive probes and used to further scan the genomic library for nearby sequences. By repeating this cycle several times, a process known as chromosome walking revealed a region around the adult hemoglobin gene that contained several closely related genes associated with hemoglobin.
6. Frequently, alleles of the same gene can be distinguished by restriction site differences in the genes themselves or in nearby locations. Alleles identified in this way are said to show restriction fragment length polymorphism. The allele responsible for sickle-cell disease was identified in this way.
7. The cystic fibrosis gene has been mapped by chromosome walking and jumping, a newer approach in which the relevant probes contain segments of the genome that are normally located about 500 kbp from one another. A cDNA library was made from normal sweat gland tissue, chosen because of the disease's association with abnormal release of sweat salt suggested that the sweat gland would contain an abundance of the messenger associated with the gene. By hybridizing the genomic DNA probes with the cDNA sweat gland library, a segment of genome was identified as a candidate for the cystic fibrosis gene. This gene was characterized in detail and found to encode a complex transmembrane protein that carries a specific amino acid change in over half of the persons with cystic fibrosis. This correlation is overwhelming support that the gene responsible for cystic fibrosis has been mapped and characterized.

Problems

1. Read the rest of the sequence in the autoradiogram in figure 27.1d as far as possible.

2. What are the major advantages of the polymerase chain reaction (PCR) method for amplifying defined segments of DNA as opposed to the use of conventional cloning methods? How might the PCR method be used to test for infection with the AIDS virus and how would this be an improvement over the antibody test currently used? (The current ELISA test is an indirect test for the presence of antibodies against the HIV proteins.)

3. Calculate the frequency of occurrence of restriction sites for PstI and HindIII in the DNA from a thermophile (80% G + C) and from *E. coli* (52% G + C).

4. You just isolated a novel recombinant clone and purified the desired insert (a 10,000 bp linear duplex DNA) from the vector. Now you wish to map the recognition sequences for restriction endonucleases A and B. You cleave the DNA with these enzymes and fractionate the digestion products according to size by agarose gel electrophoresis. Comparison of the pattern of DNA fragments with marker DNAs of known sizes yields the following results:
   a. Digestion with A alone gives two fragments, of lengths 3,000 and 7,000 bp.
   b. Digestion with B alone generates three fragments, of lengths 500, 1,000, and 8,500 bp.
   c. Digestion with A and B together gives four fragments, of lengths 500, 1,000, 2,000 and 6,500 bp.
   Draw a restriction map of the insert, showing the relative positions of the cleavage sites with respect to one another.

5. Draw the ends of a DNA fragment digested with the restriction endonuclease *Bam*HI. How do these ends differ from those generated by *Mbo*I? If *Mbo*I and *Bam*HI ends were to be ligated together, would the resulting junction be cleavable by *Bam*HI or *Mbo*I?

6. Describe a procedure for cloning a DNA fragment into the *Bam*HI site of pBR322.

7. How large a genomic library should you construct in order to detect and isolate a 15-kb gene out of a genome containing 3×10^9 bp?

8. If you were interested in isolating a cDNA for human serum albumin, why would you use a cDNA library established from mRNA isolated from liver? If you wanted to isolate the gene for albumin, why would you use a genomic library established from any human tissue?

9. Which of the *E. coli* vectors on the left (a, b, c) would be used to achieve the cloning objectives on the right (1–5)?

a. Plasmid	(1) Genomic library
b. Cosmid	(2) DNA sequencing
c. Lambda	(3) cDNA library
	(4) Small inserts
	(5) Genomic walking

10. Site-directed mutagenesis is one of the most powerful tools available to the biochemist. What are some of the applications of this technique? How can the PCR method be used to do site-directed mutagenesis, and what is the advantage of this method?

11. The Southern blot technique is often used to compare genes from different organisms. For example, one could use the human globin gene probe described in the text to determine the extent of homology between globin genes from different primates. How could one reduce the stringency of the hybridization conditions (step 4 of fig. 27.14) to permit such a "heterologous hybridization?

12. An unusual feature of the sickle-cell variant of the β-globin gene is that it directly alters a cleavage site for restriction endonuclease *Mst*II. *Mst*II recognizes the sequence CCTGAGG, which is mutated to CCTGTGG in the sickle-cell gene. How would you use this information and the Southern blot method to analyze fetal cells in amniotic fluid to determine whether the fetus carries sickle-cell anemia? What problems might you encounter in using this method?

13. Describe the procedure called "chromosome jumping." How was this procedure used to map the cystic fibrosis gene?

14. Describe a procedure using the PCR technique that could be used to determine whether a normal individual is a carrier of the cystic fibrosis ΔF_{508} mutation. What problems could you anticipate with this method?

Solutions

1. Spaces are introduced every 10 nucleotides for clarity:

5′-CAAAAAACGG ACCGGGTGTA CAACTTTTAC TATGGCGTGA CACCTAAATT
ATAGGCAGAA ATAAGTACAT GACTATTGGG AGGAGCAGGA ACAAGTAGG-3′.

The farther up the gel a fragment is, the bigger it is, and the smaller the difference between it and its neighbors. Consequently, it becomes more and more difficult to read the sequence as one gets farther and farther away from the sequencing primer.

The sequence indicated at the 5′-end would be closest to the sequencing primer. The primer sequence cannot be read from the sequencing gel, because it is not labeled. By convention (and because it makes it easier to read), the DNA sequence deduced from the sequencing gel is that of the DNA synthesized during the sequencing reactions. It represents the strand complementary to the template used by the polymerase during the sequencing reactions. The template sequence is, of course, easily deduced by writing the sequence complementary to that obtained directly from the sequencing gel. The complete sequence, with spaces introduced every 10 nucleotides for clarity, is:

5′-CAAAAAACGG ACCGGGTGTA CAACTTTTAC TATGGCGTGA CACCTAAATT
3′-GTTTTTTGCC TGGCCCACAT GTTGAAAATG ATACCGCACT GTGGATTTAA

ATAGGCAGAA ATAAGTACAT GACTATTGGG AGGAGCAGGA ACAAGTAGG-3′
TATCCGTCTT TATTCATGTA CTGATAACCC TCCTCGTCCT TGTTCATCC-5′

Observant students will notice that some bands on the sequencing gel are less intense than neighboring ones. These are most likely due to substrate and template-dependent variations in the efficiency of replication by the DNA polymerase used.

3. Because DNA is not a regularly repeating polymeric sequence, it is impossible to predict *exactly* where and how often a given restriction enzyme cleavage site will occur. However, we can predict the average probability of occurrence of a certain sequence of bases, and can therefore predict the average frequency of occurrence of the enzyme's cleavage site in any particular DNA.

   With a G + C content of 80%, and since [G] = [C] and [A] = [T], the thermophile has 40% G, 40% C, 10% A and 10% T. To calculate the probability of occurrence of a sequence of six nucleotides, multiply the probability of occurrence of each individual nucleotide. For example, the odds of the sequence -CTGCAG- (a PstI site) occurring in DNA that is 80% G + C is: $0.4 \times 0.1 \times 0.4 \times 0.4 \times 0.1 \times 0.4$, or $(0.4)^4 \times (0.1)^2$ = once every 3906 bp. An AAGCTT (HindIII) site should occur $(0.4)^2 \times (0.1)^4$ = once every 62,500 bp. Note that this DNA, with its very G + C rich composition, is predicted to be cleaved more often by the restriction enzyme with a more G + C rich recognition sequence (PstI) than by the restriction enzyme with a more A + T rich recognition sequence (HindIII).

   The *E. coli* genome composition is approximately 26% G, 26% C, 24% A, and 24% T. PstI would cleave this genome $(0.26)^4 \times (0.24)^2$ = once every 3,800 bp, and HindIII, $(0.26)^2 \times (0.24)^4$ = once every 4,500 bp. Note that because the *E. coli* nucleotide composition is not as strongly biased as that of the thermophile, PstI and HindIII sites are expected to occur with similar frequencies.

   Typically, of course, there is considerable variability in the sizes of restriction fragments produced by any one enzyme on a given DNA. The above estimates represent probable averages, and the range of restriction fragment sizes can still vary from a few base-pairs to millions of base-pairs.

5. There is no difference between the single-stranded ends generated by BamHI and by MboI. These ends are complementary (they are termed "compatible" ends) and they can anneal and be ligated together by DNA ligase, as shown below:

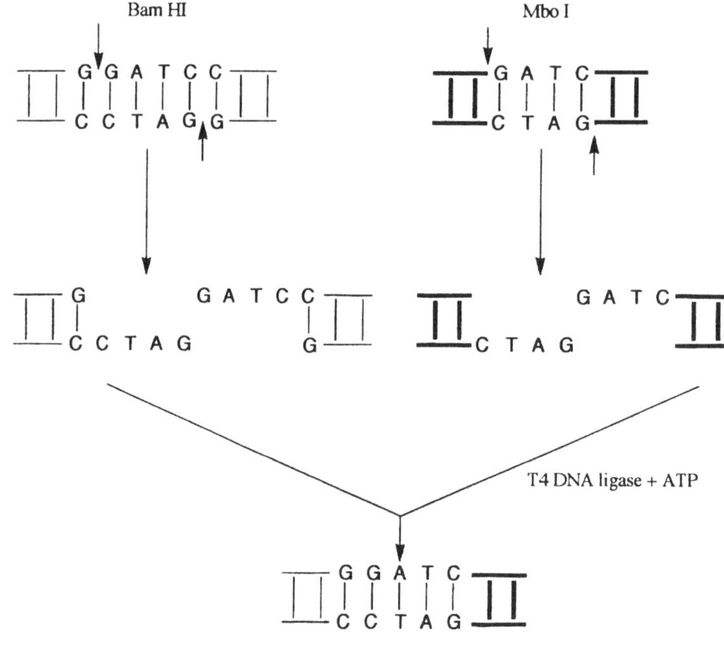

The resulting sequence contains an MboI site and can be cleaved by that enzyme. It only has a 25% probability of restoring a BamHI site, however, depending on the nucleotide located adjacent to the MboI site. In other words, if the nucleotide 5′ to the MboI site is a G, ligation of MboI to BamHI would regenerate a sequence cleavable by BamHI, but not if it is an A, C, or T.

7. In order to isolate a 15-kb gene from a genome containing 3×10^9 bp, it would be necessary to isolate $3 \times 10^9/15{,}000$ or 200,000 fragments per genome, or 200,000 clones. Table 27.2 in the text describes for several examples the number of clones required to represent the entire genome of various organisms, assuming DNA fragments of a given size. It is recommended that a library 3 to 10 times the minimum size should be prepared, to ensure a high probability that a given fragment will be represented at least once. In the case of the gene specified, the library should therefore contain between 6×10^5 and 2×10^6 clones.

9. a. 2, 4, sometimes 3
   b. 1, 5
   c. 1, 3

11. To reduce the hybridization stringency, researchers choose conditions which stabilize double-stranded DNA, allowing sequences which share only partial complementarity to form base-pairs. Examples of such conditions include reducing the hybridization temperature, and increasing the salt concentration in solution. A reduction in temperature helps stabilize the duplex formed between heterologous DNAs, which have a lower melting temperature than perfectly paired DNA. The salt helps reduce the repulsion between phosphodiester backbones, again facilitating the formation of base-paired regions.

13. Chromosome jumping is a useful procedure to traverse long distances and skip troublesome regions of the genome (repetitious sequences, etc.). One approach with this technique is to digest the genome with rare cutting restriction enzymes (*Not* I recognizes an 8-base sequence and generates an average fragment size of 500 kb of DNA). The fragments are circularized with a small marker DNA between the ends. The marker DNA contains sequences necessary for cloning in λ. These circular DNA fragments are then cleaved with another restriction enzyme that produces fragments small enough to clone in λ. With this procedure only clones that contain the marker DNA, i.e., the ends of the original fragment, are isolated. This method would permit only one jump; thus another method used with this technique is called a linking library. The same sample of DNA is digested with a restriction enzyme that gives smaller fragments. These smaller fragments are then circularized with the same marker DNA used in the jumping library. The circular DNA is then digested with *Not* I to linearize the fragments for insertion into the vector. The linking library carries sequences on both sides of the *Not* I restriction sites while the jumping library carries sequences from one side of two adjacent *Not* I sites. These two libraries are then used as shown in the figure.

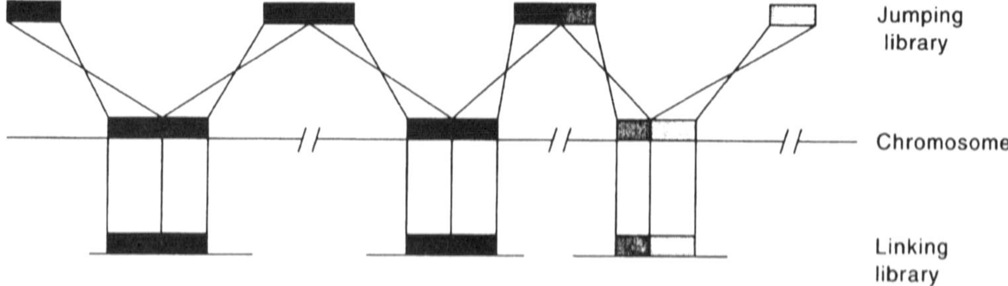

28 RNA Structures and Metabolism

Summary

In this chapter we described the synthesis, transcription, and posttranscriptional reactions undergone by the three major classes of RNA. The main points we covered are as follows:

1. RNA is synthesized in the 5′ → 3′ direction by the formation of 3′-5′-phosphodiester linkages between four ribonucleoside triphosphate substrates, analogous to the process of DNA synthesis. The sequence of bases in RNA transcripts assembled by DNA-dependent RNA polymerases is specified by the complementary sequences of the DNA template strand.
2. Some newly synthesized RNA transcripts are the functional species, whereas others must be modified or processed into the mature functional species. Modifying enzymes add nucleotides to the 5′ or 3′ ends or alter bases within the RNA, such as by methylation of specific residues. Specific processing enzymes cleave RNA internally, splice together noncontiguous regions of a transcript, or remove nucleotides from the 5′ or 3′ ends.
3. The major classes of RNA in both prokaryotes and eukaryotes are messenger RNA, ribosomal RNA, and transfer RNA. These distinct classes of RNA play specific functional or structural roles in the translation of genetic information into proteins.
4. DNA-dependent synthesis of RNA in *E. coli* is catalyzed by one enzyme, consisting of five polypeptide subunits. The complete holoenzyme is composed of four polypeptides (the core enzyme) and an additional polypeptide that confers specificity for initiation at promoter sequences in the DNA template.
5. The steps involved in transcription include binding of polymerase at the initiation site, initiation, elongation, and termination.
6. In eukaryotes most transcription takes place in the nucleus. Three nuclear RNA polymerases, I, II and III, are responsible for the synthesis of rRNA, mRNA, and small RNA transcripts, respectively. The polymerases contain more subunits than in *E. coli*, and other proteins must bind at the initiation sites or near the initiation sites before the polymerases can begin transcription.
7. Viruses sometimes use the host RNA polymerase in a modified form and sometimes synthesize their own RNA polymerase.
8. One of the more interesting processing reactions of nascent transcripts involves the removal of internal sequences. This type of processing is referred to as splicing. Most splicing reactions appear to require host proteins; however, some only require RNA, a fact demonstrating that RNA is capable of functioning like an enzyme in making and breaking of phosphodiester linkages in polyribonucleotides. Catalytic RNAs may have evolutionary significance.
9. Inhibitors of RNA synthesis may be classified according to their mechanism of action. Some bind to DNA, some bind to RNA, and some are incorporated into the growing RNA chain during transcription.

Problems

1. Compare the reactions catalyzed by RNA and DNA polymerases. What are the similarities and differences?

2. One strand of DNA is completely transcribed into RNA by RNA polymerase. The base composition of the DNA template strand is: G = 20%, C = 25%, A = 15%, T = 40%. What would you expect the base composition of the newly synthesized RNA to be?

3. Although 40 to 50% of the RNA being synthesized in *E. coli* at any given time is mRNA, only about 3% of the total RNA in the cell is mRNA. Explain.

4. In *E. coli*, the mRNA fraction is heterogeneous in size, ranging from 500 to 6,000 nucleotides. The largest mRNAs have many more nucleotides than needed to make the largest proteins. Why are some of these mRNAs so large in bacteria?

5. Base-stacking interactions are an important stabilizing force in nucleic acid structures. Describe how base stacking contributes to the tertiary structure of the tRNA molecule.

6. Why is a single-strand binding protein or a DNA helicase not required for transcription as they are for replication?

7. When yeast phenylalanine tRNA is digested with a small amount of RNase (partial digest) such as T2 RNase, almost all of the cleavage sites are found in the anticodon loop. Explain this observation.

8. Speculate on the advantage of having three rRNAs (16S, 23S, and 5S) as part of the same RNA precursor (such as is found in *E. coli*)?

9. In *E. coli* the precise spacing between the –35 and –10 conserved promoter elements has been found to be a critical determinant of promoter strength. What does this suggest about the interaction between RNA polymerase and these elements? What sorts of evidence could you obtain about this interaction by doing "footprint" experiments? (Also explain how you would do these experiments.)

10. Human RNA polymerase II generates RNA at a rate of approximately 3,000 nucleotides per minute at 37°C. One of the largest mammalian genes known is the 2,000-kbp (1 kbp = one kilo base-pair = 1,000 base-pairs) gene encoding the muscle protein, dystrophin. How long would it take one RNA polymerase II molecule to completely transcribe this gene? How long would it take to transcribe the adult β-globin gene (1.6 kbp)?

11. In early research with intact eukaryotic mRNA, the RNA appeared to have two 3′-ends and no 5′-terminus. Explain these observations.

12. What is unique about the promoter for RNA polymerase III? Diagram how this promoter works.

13. It has recently been demonstrated that the TATA-binding protein, in addition to flattening and widening the DNA by interacting (atypically) with the minor groove, also introduces a sharp bend (> 100°) into its recognition sequence on binding. How might these changes in the structure of the DNA facilitate assembly of the RNA polymerase II transcription initiation complex?

14. Propose a sequence for the guide RNA that directs the editing of precursor RNA to yield the mRNA shown in figure 28.20.

15. A particular eukaryotic DNA virus is found to code for two mRNA transcripts, one shorter than the other, from the same region on the DNA. Analysis of the translation products reveals that the two polypeptides share the same amino acid sequence at their amino-terminal ends but are different at their carboxyl-terminal ends. The longer polypeptide is coded by the shorter mRNA! Suggest an explanation.

16. Although a phosphodiester bond must be formed between the upstream and downstream exons, there is no direct ATP requirement for splicing of mRNA introns or for splicing of the *Tetrahymena* pre-rRNA intron. Explain this observation.

Solutions

1. RNA and DNA polymerases catalyze the same overall reaction mechanistically, involving hydrolysis of a nucleotide triphosphate to release pyrophosphate and form a phospodiester bond. In both cases, the order of nucleotide addition is specified by the template, and synthesis of the growing nucleic acid chain is in a 5′ to 3′ direction (the enzymes move in a 3′ to 5′ direction along the template strand).

   In addition to the obvious difference in substrates (RNA polymerase utilizes ribonucleotides while DNA polymerase utilizes deoxyribonucleotides), these two enzymes differ in their requirements for initiating synthesis: RNA polymerase initiates *de novo*, and does not require a primer, unlike DNA polymerase. Transcription does not generally require a separate DNA helicase enzyme because it is not necessary to separate the two strands of DNA along their entire length. Finally, RNA polymerase does not have any proof-reading activity, unlike DNA polymerases, which generally have a 3′ to 5′ exonuclease activity to remove misincorporated nucleotides.

3. mRNA in bacteria is degraded extremely rapidly, with a turnover rate of approximately 50% in 1 to 3 min, and thus does not accumulate. The 3% of the total RNA present at any time represents that mRNA which has just been synthesized but not yet degraded. However, rRNA and tRNA are considerably more stable and accumulate. Even though these RNAs together account for roughly 50% of the total RNA synthesis, they represent greater than 95% of the total RNA present at any given time.
5. The bases within the base-paired region of each arm of the tRNA cloverleaf stack in a similar fashion to the base stacking described in Chapter 25 for DNA. (There is an important difference between the RNA double helix and B-DNA in that the 2′-OH of RNA cannot fit in a B-form helix, and the base-paired regions in tRNA and other RNA molecules adopt a conformation known as the A-form helix, in which the bases are more tilted.) In addition to the base stacking within base-paired regions, there is also stacking of one helix on top of another in the tRNA molecule. In particular, the acceptor stem stacks with the TΨC stem and loop to form one nearly continuous stacked double helix. The anticodon stem and the D stem also stack on top of one another (see fig. 28.3).
7. If yeast phenylalanine tRNA was partially digested with a RNase, the loops or single-stranded regions would be cleaved. The cloverleaf model for the secondary structure of tRNAs predicts that cleavages would occur in the D and T loops as well as the anticodon loop. However, the D and T loops interact with each other in the three-dimensional model of tRNA, leaving only the anticodon loop exposed to RNase cleavage.
9. The RNA polymerase binds to the same side of the duplex at the –10 and –35 regions with about two turns of the duplex helix between these boxes. This binding site can be determined by a variety of "footprint" experiments. DNA that is 5′-labeled can be mixed with the polymerase, and regions that are protected from digestion with an enzyme like DNase I can be determined on a sequencing gel. Sequences with tight contact with the RNA polymerase are observed as a series of blank spots in the sequencing ladder (see Methods of Biochemical Analysis 28A in the text) that look like footprints.
11. One common way to determine the 3′-end of an RNA is to treat the RNA with Na-periodate and oxidize the cis-diols to dialdehydes, which can then be reduced to dialcohols with $^3H\text{-}BH_4$. Normally the only nucleotide in the sequence that has cis-diols is the 3′-terminal, which has the 2′ and 3′ hydroxyls. However, since the eukaryotic mRNAs are capped at the 5′-end (GpppGpNpN—N_{OH}) they also have a guanosine at the 5′-terminus as well as the adenosine (most eukaryotic mRNAs end in poly A) at the 3′-end with cis-diols. Also, these eukaryotic mRNAs would appear to lack a normal 5′-end. Many types of 5′-end analysis would not detect the 5′-terminus, leading to the paradox of a RNA with no 5′-end and two 3′-ends.
13. By flattening and widening the DNA minor groove, TATA binding protein may assist other factors to form an open complex-like structure with separated strands. The bending induced by TATA binding protein would bring the DNA upstream and downstream of the TATA box closer together, promoting interactions between proteins bound to upstream elements and those near the start site of transcription.
15. If an intron or part of an intron containing a stop signal for translation was not removed, this mRNA would be longer but would yield a shorter polypeptide. Alternative splicing would remove the intron or use an alternative splice site, generating a shorter mRNA and a longer polypeptide.

29 Protein Synthesis, Targeting and Turnover

Summary

We focused in this chapter on the complex mechanisms of protein synthesis. The following points are central to this subject.

1. Three types of RNA carry out protein synthesis: Ribosomal RNA, transfer RNA, and messenger RNA. Ribosomal RNA is invariably complexed with many proteins to form ribosomes, on which amino acids are assembled into polypeptides. The amino acids are brought to the ribosomes attached to transfer RNAs. The messenger RNA contains the instructions for translation in the form of the genetic code. Messenger RNAs form transient complexes with ribosomes. Individual aminoacyl-tRNAs bind to specific sites on the messenger RNAs. The interacting site on the messenger is the codon; the interacting site on the tRNA is the anticodon.
2. The part of the messenger that is translated is the reading frame. Eukaryotic messages carry only one reading frame, whereas prokaryotic messengers may carry more than one. In prokaryotes the initiation codons are recognized by a ribosome-binding site upstream of the start codon.
3. Most transfer RNAs have common parts and uncommon parts. The common parts facilitate binding of the aminoacyl-tRNAs to common sites on the ribosome. The uncommon sites permit specific reactions with charging enzymes that covalently attach the correct amino acids to the correct tRNA. Another uncommon site on the tRNAs is the anticodon, which leads to specific complex formation with the complementary codon site on the messenger.
4. Attachment of the amino acid to the tRNA is catalyzed by a specific aminoacyl synthase, which recognizes all the cognate tRNAs for a specific amino acid.
5. A unique methionyl-tRNA binds to the initiation codon on all messages.
6. The genetic code is the sequence relationship between nucleotides in the messenger RNA and amino acids in the proteins they encode. Triplet codons are arranged on the messenger in a nonoverlapping manner without spacers.
7. The code was deciphered with the help of synthetic messengers with a defined sequence, by analyzing the types of polypeptide chains that were made when these messengers were used in an *in vitro* protein-synthesizing system.
8. The genetic code is highly degenerate, with most amino acids represented by more than one codon. In many cases the 3′ base in the codon may be altered without changing the amino acid that is encoded.
9. The codon-anticodon interaction is limited to Watson-Crick pairing for the first two bases in the codon but is considerably more flexible in the third position.
10. Translation begins with the binding of the ribosome to mRNA. A number of protein factors transiently associate with the ribosome during different phases of translation: Initiation factors, elongation factors, and termination factors.
11. Initiation factors contribute to the ribosome complex with the messenger RNA and the initiator methionyl-tRNA. Elongation factors assist the binding of all the other tRNAs and the translocation reaction that must occur after each peptide bond is made. Termination factors recognize a stop signal and lead to the termination of polypeptide synthesis and the release of the polypeptide chain and the messenger from the ribosome.

12. A large number of antibiotics have been characterized that inhibit protein synthesis. These antibiotics are usually made by a particular microorganism, and they inhibit protein synthesis in a broad family of other organisms, mostly bacterial.
13. Specific enzymes catalyze folding after polypeptide synthesis.
14. Proteins are targeted to their destination by signal sequences built into the polypeptide chain. These signals are usually located at the N-terminal end of the protein and are generally cleaved during protein maturation.
15. Posttranslational modifications include many covalent alterations: Polypeptide processing, attachment of carbohydrate or lipid groups to specific side chains, and addition of many other low-molecular-weight ligands to side chains.
16. Intracellular protein degradation is not random. Different proteins have quite different half-lives, which are related to specific structural features. Imperfectly folded proteins and polypeptide fragments are frequently degraded most rapidly. In eukaryotes, lysosomes play a major role in protein degradation.

Problems

1. Compare the translation initiation signals in prokaryotic and eukaryotic systems, and describe those features of each type of mRNA that determine the frequency with which a particular message is translated. What consequences do these differences have for gene organization in the two systems?

2. The relationship between tRNAs and their synthases is sometimes called the "second genetic code." Explain.

3. A single tRNA can insert serine in response to three different codons: UCC, UCU, or UCA. What is the anticodon sequence of this tRNA?

4. How much energy is required to synthesize a single peptide bond in protein synthesis? How does this compare with the free energy of formation of the peptide linkage, which is about 5 kcal/mole?

5. Explain this statement: "The universal genetic code is not quite universal."

6. Explain why the use of GUG and UUG as initiation codons in place of AUG was not expected, even based on Crick's wobble hypothesis.

7. Assuming that translation begins at the first codon, deduce the amino acid sequence of the polypeptide encoded by the following mRNA template:

   AUGGUCGAAAUUCGGGACACCCAUUUGAAGAAACAGAUAGCUUUCUAGUAA

8. Assume that you have a copolymer with a random sequence containing equimolar amounts of A and U. What amino acids would be incorporated and in what ratio, when this copolymer is used as an mRNA?

9. Researchers often design degenerate oligonucleotides based on a protein sequence for use as hybridization probes to isolate the corresponding gene. (A degenerate oligonucleotide is actually a mixture of oligonucleotides the sequences of which differ at positions corresponding to degeneracies in the genetic code.) The N-terminal amino acid sequence of a protein is:

   Met-Val-Asp-Ser-Asn-Trp-Ala-Gln-Cys-Asp-Pro-Ala-Thr

   Give the sequence of the least degenerate 20-residue-long oligonucleotide that hybridizes to the gene encoding this protein.

10. The effect of single-point mutations on the amino acid sequence of a protein can provide precise identification of the codon used to specify a particular residue. Assuming a single base change for each step, deduce the wild-type codon in each of the following cases.

a. Gln ⟶ Arg ⟶ Trp

b. Glu ⟶ Lys ⟶ Ile

c.

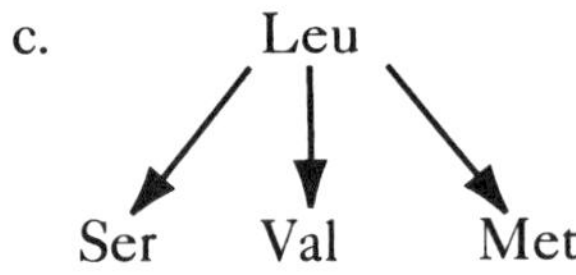

d. Thr ⟶ Ile, Pro, Lys

11. Even though the roles of IF-2, EF-Tu, EF-G, and RF-3 in protein synthesis are quite different, all four of these proteins share a domain with significant amino acid sequence similarity. Suggest a role for this conserved domain.

12. The antibiotic fusidic acid inhibits protein synthesis by preventing EF-G from cycling off of the ribosome. Fusidic-acid-resistant mutants of EF-G have been isolated. Fusidic acid resistance is recessive to sensitivity. In other words, an *E. coli* cell containing two EF-G genes, one resistant and one sensitive, is still sensitive to the antibiotic. Why? (*Hint:* Look at fig. 29.2.)

13. What are the major differences between mechanisms of protein import into endoplasmic reticulum compared to protein import into mitochondria?

14. Scientists have tried to isolate the peptidyl transferase from ribosomes for many years without success. It is now thought that this activity is part of the ribosome (large subunit). Discuss this point, in view of what you know about other catalytic RNP complexes (RNA-protein complexes).

15. What are the possible amino acid changes that can result from a single nucleotide change in a GAA codon? Knowing the structures of the amino acids, what do you predict are the effects of the altered amino acids?

Solutions

1. In prokaryotes, translation initiation requires both an AUG triplet, or initiation codon, and a purine-rich element, termed the Shine-Dalgarno sequence, located approximately 10 bases upstream of the AUG. The initiation codon is not generally the most 5′ proximal AUG in the mRNA. The Shine-Dalgarno sequence is thought to base-pair with a pyrimidine-rich sequence located toward the 3′ end of the 16S ribosomal RNA of the 30S ribosomal subunit, which helps align the AUG triplet for initiation. The efficiency with which a given mRNA is translated is determined by the homology of its Shine-Dalgarno sequence to the consensus (i.e., the extent of base pairing between the mRNA and 16S rRNA), and the sequence context of the initiation codon.

   In eukaryotes, translation generally begins at the AUG triplet that lies closest to the 5′ end of the mRNA. This is thought to be recognized by a scanning mechanism, in which the 40S ribosomal subunit first binds to the mRNA at the capped 5′ end, and then migrates 5′ → 3′ until it encounters the first AUG codon. (There is as yet no evidence to support the possibility of the base pairing between 18S rNA and the 5′ end of eukaryotic mRNA in initiation.) The efficiency with which a given AUG is recognized as a translation start site is determined by its sequence context; CCACC*AUG*G is the optimal context.

   The difference in the mechanism of translation initiation in prokaryotes compared to eukaryotes has profound consequences for the strategy used to coordinate expression of a set of genes in the two systems. In prokaryotes this coordination is achieved by organizing genes into transcription units which are transcribed to give polycistronic mRNAs. Each cistron within a polycistronic mRNA begins with a Shine-Dalgarno sequence and initiation AUG codon. In contrast, in eukaryotes, in which translation begins almost invariably (and exclusively) at the 5′ most AUG codon, monocistronic mRNAs are necessarily the order of the day. Genes which must be coordinately expressed are consequently not organized into transcriptional units, and coordination must be achieved in some other way.

3. This serine tRNA probably has inosine at the 5′ position of its anticodon, which can pair (based on the wobble rules) with U, C, or A. Thus the anticodon of this tRNA is most likely IGA (given in the correct 5′ to 3′ direction). This tRNA can recognize the three different codons as follows:

tRNA Anticodon:	3′-AGI-5′	3′-AGI-5′	3′-AGI-5′
	| | |	| | |	| | |
Codon:	5′-UCA-3′	5′-UCC-3′	5′-UCU-3′

See figure 29.8 for the base-pair structure between inosine and A or U. The I-C base-pair is almost identical to a G-C base-pair, except that a single hydrogen bond is missing. (I is just like G except that it lacks the exocyclic amino group at position 2.)

5. A number of variations have been found in the universal genetic code in genes found in mitochondria. These variations in the meaning of some of the code words represent divergences from the standard genetic code and not an independent origin of another genetic code. These divergences probably arose in these organelles because of the limited number of genes coded and requirements for the synthesis of ribosomes and tRNA. It was clear that something was unusual when only 24 types of tRNAs were found in mitochondria. Mitochondria do not use all 61 codons (see table 29.3 in the text for the yeast mitochondria genetic code).

7. On the basis of the information provided in table 29.1, the amino acid sequence corresponding to the template given may be deduced as follows:

AUG	GUC	GAA	AUU	CGG	GAC	ACC	CAU	UUG	AAG	AAA	CAG	AUA	GCU	UUC	UAG	UAA
Met	Val	Glu	Ile	Arg	Asp	Thr	His	Leu	Lys	Lys	Gln	Ile	Ala	Phe	Ter	Ter

9. To minimize the degeneracy of the oligonucleotide to be synthesized, the amino acid sequence should first be examined for those amino acids which are encoded by the smallest number of codons in the universal genetic code.

	Met-	Val-	Asp-	Ser-	Asn-	Trp-	Ala-	Gln-	Cys-	Asp-	Pro-	Ala-	Thr
# of codons:	1	4	2	6	2	1	4	2	2	2	4	4	4

The seven amino acids from asparagine to proline should be used to design the oligonucleotide with the lowest degeneracy. An instrument called a DNA synthesizer can be programmed to synthesize a mixed oligonucleotide with mixtures of nucleotides at positions corresponding to the wobble position in the genetic code. With spaces introduced to facilitate reading, the sequence corresponding to the above amino acids would be: (5′)AAY TGG GCN CAR TGY GAY CC(3′). R is the abbreviation for a mixture of A or G (R = pu*R*ine), Y is the abbreviation for a mixture of C and T (Y = p*Y*rimidine), and N stands for a mixture of all four *N*ucleotides. AAY is the DNA equivalent of the two asparagine codons (AAU and AAC) in the universal genetic code, etc. This oligonucleotide would be expected to hybridize to the sequenced protein's gene (but not its mRNA).

The complementary oligonucleotide, GGRTCRCAYTGNGCCCARTT, would also be a correct answer for the question as stated. (This sequence was derived by writing the complementary nucleotides in the correct 5′ to 3′ direction.) The complementary oligonucleotide might be desirable for some experiments because it should hybridize to both DNA and mRNA corresponding to this protein.

The oligonucleotide sequences given are $2 \times 1 \times 4 \times 2 \times 2 \times 2$ = 64-fold degenerate. In other words, there are 64 different sequence combinations in this mixture of oligonucleotides. Note that oligonucleotides based on amino acid sequences typically are of the length [3n – 1] to avoid the degenerate 3′-most nucleotide of the last amino acid codon. In the example given above, the addition of one extra nucleotide at either end would increase the degeneracy of the oligonucleotide four-fold.

11. Although IF-2, EF-Tu, EF-G and RF-3 play different roles in protein synthesis, they all have in common the ability to bind and hydrolyze GTP, and the ability to interact with bacterial ribosomes. The X-ray crystal structure of EF-Tu bound to GDP and GTP is known. Based on this structure, it appears that many of the amino acids that are shared between IF-2, EF-Tu, EF-G and RF-3 are somehow involved in interactions with GTP and GDP. Thus, these proteins

(and many others, including the proto-oncogene, Ras) share similar amino acids within the GTP-binding domain. A few of the conserved amino acids may also be implicated in ribosome binding, suggesting that the GTP-binding protein synthesis factors all interact with some of the same residues of the ribosome. However, this aspect of elongation factor function is not yet as well-documented as the binding of guanine nucleotides.

13. Import into the endoplasmic reticulum requires an N-terminal signal sequence which contains a long stretch of hydrophobic amino acids. In contrast, mitochondrial N-terminal transit peptides are hydrophilic, and rich in serine and threonine, with regularly spaced basic amino acids. Import into the ER requires the signal recognition particle (SRP) and its receptor, while mitochondrial import does not require the SRP and presumably uses a different receptor. Synthesis of proteins targeted to the ER occurs on polysomes attached to the ER, while proteins destined for mitochondrial import are synthesized in the cytoplasm. Import into mitochondria requires ATP and a membrane potential. Import into the ER requires GTP in the initial docking step.

15. GAA encodes a glutamic acid residue. If this glu is essential for the catalytic activity of the protein, any mutation (except GAA to GAG) will be deleterious to the activity of the protein. If the glu residue is not essential for catalysis or proper folding, the possible mutations, in order of increasing potential severity are:

GAA changes to:	*Glu changes to:*	*Likely effect:*
GAG	Glu	None
GAT or GAC	Asp	Very little effect, because asp has similar acidic side-chain
CAA	Gln	Very little effect, because gln is just amide of glu
GCA	Ala	Hard to predict, but may not have much of an effect, because ala has a small side-chain, and like glu, tends to form α-helix structure
AAA	Lys	May have significant effect on enzyme activity because it is a basic amino acid in place of an acidic one
GTA	Val	May have significant effect on enzyme activity because it is uncharged, and, unlike glu, prefers to be in β-sheet
GGA	Gly	Will probably affect enzyme activity, because it is conformationally flexible, and tends to form β bends
TAA	Terminator	Almost certain to disrupt expression of the protein

It is clear from the above analysis that many potential mutations of a GAA codon might have little or no effect on the encoded protein. The genetic code seems to have evolved to minimize the effects of many mutations, *i.e.*, similar amino acids are often found in rows and columns in the universal genetic code.

In some cases, the mutations described above might affect expression of the protein, by affecting the translatability of the mRNA or protein stability. These sorts of effects would be difficult to predict.

30 Regulation of Gene Expression in Prokaryotes

Summary

In this chapter we discussed the regulatory systems of the *E. coli* bacterium and the λ bacteriophage. The main points in our presentation are as follows.

1. *Escherichia coli* carries about 3,000 genes. Only a small fraction of the genome is actively transcribed at any given time. But all of the genes are in a state where they can be readily turned on or turned off in a reversible fashion. The level of transcription is regulated by a complex hierarchy of control elements.
2. In the most common form of control, expression is regulated at the initiation site of transcription. There are several ways of doing this, all of them revolving around protein or small-molecule factors that influence the binding of RNA polymerase at the transcription start site.
3. The *lac* operon, a cluster of three genes involved in the catabolism of lactose, exemplifies both positive and negative forms of control that influence the rate of initiation of transcription. Jacob and Monod identified the repressor as a negative control element that is *trans*-dominant and the operator as a *cis*-dominant site for binding the repressor. Transcription is initiated by RNA polymerase at the promoter, which overlaps the operator site on one side of the three structural genes of the *lac* operon. The tight complex between repressor and operator prevents initiation, and it is broken when lactose is present. The lactose is readily converted to allolactose, which binds to the lac repressor. This changes the structure of the repressor so that it dissociates from the DNA.
4. Initiation of transcription proceeds at a greatly increased rate when cyclic AMP is present. This is because cyclic AMP forms a complex with the CAP activator protein, which then binds at a site adjacent to the polymerase-binding site. The CAP protein enhances polymerase binding at the adjacent site by cooperative binding.
5. Lactose is the substrate of the enzymes of the *lac* operon. In the absence of lactose, there is no use for enzymes of the *lac* operon.
6. CAP and cAMP activate a large number of genes in *E. coli* that are concerned with catabolism. When glucose is present, the cAMP is greatly lowered and the *lac* operon is expressed at a very low level, even when lactose is present. This is because glucose is a more readily metabolizable carbon source than lactose.
7. The *trp* operon contains a cluster of five structural genes associated with tryptophan biosynthesis. Initiation of transcription of the *trp* operon is regulated by a repressor protein that functions similarly to the lac repressor. The main difference is that the trp repressor action is subject to control by the small-molecule effector, tryptophan. When tryptophan binds the repressor, the repressor binds to the *trp* operator. Thus, the effect of the small-molecule effector here is opposite to its effect on the *lac* operon. When tryptophan is present, there is no need for the enzymes that synthesize tryptophan.
8. The *trp* operon has a control locus called an attenuator about 150 bases after the transcription initiation site. The attenuator is regulated by the level of charged tryptophan tRNA, so that between 10% and 90% of the elongating RNA polymerases transcribe through this site to the end of the operon. Low levels of *trp* tRNA encourage transcription through the attenuator.

9. Ribosomal RNA and protein synthesis are both controlled at the level of initiation of transcription. This is a result of the direct binding of guanosine tetraphosphate, ppGpp, to the RNA polymerase. This binding decreases the affinity of RNA polymerase for the initiation sites of transcription. Guanosine tetraphosphate is synthesized when the general level of amino-acid-charged tRNA is low.
10. The synthesis of ribosomal proteins is regulated at the level of translation. Certain ribosomal proteins bind to specific sites on the ribosomal RNAs or their own mRNAs. In the absence of the ribosomal RNAs, they bind to their own mRNAs, which inhibits their translation. This form of translational control regulates the rate of synthesis of ribosomal proteins so that it does not exceed the rate of ribosomal RNA synthesis.
11. Viruses borrow heavily on the host enzymatic machinery to obtain energy for synthesis, as well as for replication, transcription, and translation. The virus infective cycle is strongly irreversible. Virus infection is followed by the gradual turning on of viral genes. Viral enzymes are the first viral gene products; in late infection, the virus structural proteins are favored. The irreversible lytic cycle of the virus is directed by a cascade of controls.
12. In λ the host RNA polymerase is used throughout. Regulation is achieved through a series of repressors and activators, as well as two viral proteins that bind directly to the RNA polymerase. The viral proteins that bind to the polymerase modify it so that it can transcribe through provisional stop signals.
13. When λ phage infects an *E. coli* cell, it does not always produce viral progeny. Sometimes it integrates its genome into the host genome and replicates only as the host genome replicates. This so-called lysogenic state can be disrupted by DNA-damaging conditions such as exposure to UV radiation. Under these conditions the dormant viral genome enters the active replication cycle.
14. Bacterial regulatory proteins are controlled by small-molecule effectors; viral regulatory proteins are not. Bacterial genes are regulated in a highly reversible manner; viral genes are usually turned on only once.
15. Proteins that regulate transcription usually bind to specific sites on the DNA. The recognition process involves specific hydrogen bonds formed between amino acid side chains of the protein and the base pairs of the DNA. Most of this interaction takes place in the major groove of the DNA, which is more accessible to the protein. The vast majority of regulatory proteins interact with the DNA from the side chains of a segment of α helix that fits snugly into the major groove. The binding site on the DNA usually consists of two half-sites, which are arranged on a dyad axis of symmetry that matches two half-sites on the regulatory protein. The two half-sites are situated in adjacent major grooves on one side of the DNA.

Problems

1. What set of data originally led Jacob and Monod to suggest the existence of a repressor in *lac* operon regulation?

2. The *lac* promoter—operator region is found in many vectors used by molecular biologists to clone genes. Not infrequently, high levels of transcription driven by the *lac* promoter generate toxic levels of the cloned gene product, causing *E. coli* cells to grow poorly. Faced with such a situation, how could you minimize expression of the *lac* promoter?

3. With respect to β-galactosidase production in the presence and absence of inducer, what would be the phenotype of the following *E. coli* mutants:
   a. $i^s o^+ z^+$
   b. $i^s o^c z^+$
   c. $i^s o^c z^- / i^+ o^+ z^+$
   d. $i^+ o^c z^- / i^- o^+ z^+$

4. In a cell that is *lacZ*$^-$, what would be relative thiogalactoside transacetylase concentration, compared with wild type, under the following conditions?
   a. After no treatment
   b. After addition of lactose
   c. After addition of IPTG

5. Consider a negatively controlled operon with two structural genes (*A* and *B*, for enzymes A and B), an operator gene (*O*), and a regulatory gene (*R*). The first line of data in the table gives the enzyme levels in the wild-type strain after growth in the absence or presence of the inducer. Complete the table for the other cultures.

	Uninduced		Induced	
Strains	*Enz* A	*Enz* B	*Enz* A	*Enz* B
Haploid strains				
(1) $R^+O^+A^+B^+$	1	1	100	100
(2) $R^+O^cA^+B^+$				
(3) $R^-O^+A^+B^+$				
Diploid strains				
(4) $R^+O^+A^+B^+/R^+O^+A^+B^+$				
(5) $R^+O^cA^+B^+/R^+O^+A^+B^+$				
(6) $R^+O^+A^-B^+/R^+O^+A^+B^+$				
(7) $R^-O^+A^+B^+/R^+O^+A^+B^+$				

6. Explain why, when *E. coli* is grown in the presence of *both* glucose and IPTG, *β*-galactosidase protein levels are lower than when grown only in the presence of IPTG.

7. Although *E. coli* promoters generally conform to a rather well-defined consensus sequence, no perfect match to this consensus has ever been observed in a naturally occurring promoter. Suggest an explanation.

8. Do you expect a regulatory mechanism like the prokaryotic attenuator to be found in eukaryotes? Why or why not?

9. The *lac* repressor has an "on" rate constant for the binding of the *lac* operator (when cloned into λ) of about $5 \times 10^{10} \text{M}^{-1}\text{s}^{-1}$. This value is much greater than the calculated diffusion-controlled process, which is about $10^{8}\text{M}^{-1}\text{s}^{-1}$ for a molecule the size of the lac repressor. Explain why this repressor binding works better than expected.

10. A mutation in the *trp* leader region is found to result in a reduction in the level of *trp* operon expression when the mutant is grown in rich medium. However, when the mutant is grown in a medium lacking glycine, a stimulation in the level of *trp* enzymes is observed. Explain these observations. What do you anticipate is the effect of growing the mutant in a medium lacking both glycine and tryptophan?

11. In *E. coli* no pools of free rRNAs or ribosomal proteins are floating around in the cell even when the bacteria are grown at different growth rates. Explain how *E. coli* coordinates the biosynthesis of the ribosome.

12. Draw a graph of rRNA gene transcription levels under the conditions shown in figure 30.16.

13. The rRNA genes of *E. coli* are present in multiple copies to facilitate production of many copies of rRNA during periods of rapid growth. If ribosomal proteins and RNAs need to be assembled in a 1:1 ratio, then why are single-copy genes adequate for ribosomal protein expression?

14. Describe the principal differences between patterns of control of gene expression used by bacterial host and bacteriophage systems.

15. How does the clustering of genes on the bacteriophage λ genome (fig. 30.20) facilitate its genetic regulation?

16. How is the synthesis of the CAP protein regulated? What is unusual about this regulation?

17. Gene regulatory proteins in bacteria were predicted (before their precise structure was known) to interact in the major groove of DNA by a two-site model of binding. Describe the data that showed this model to be correct.

18. Referring to figure 30.27c, draw a detailed structure of the interaction of residue Arg69 of the *trp* repressor with G9 of its recognition sequence. Explain why binding of repressor proteins to DNA does not disrupt the DNA double helix.

Solutions

1. The existence of a repressor in *lac* operon regulation was first suggested by the results of studies of merodiploids. In merodiploids of the type i^+z^-/Fi^-z^+, Jacob and Monod were able to

demonstrate that the i^+ (inducible) allele is dominant to the i^- (constitutive) allele when on the same chromosome (*cis*) or on a different chromosome (*trans*) with respect to the z^+ allele. The fact that the i^+ allele was dominant in *trans* indicated to Jacob and Monod that *i* gene mutations belong to an independent cistron that governs the expression of *z*, *y*, and *a* genes through the production of a diffusible cytoplasmic component, the lac repressor.

3. a. Cells with the genotype $i^so^+z^+$ have a "super-repressor" i^s mutation, which causes the repressor to be insensitive to inducer. Thus, even though the operator and β-galactosidase loci are wild-type, no β-galactosidase will be produced in this mutant. On media containing X-gal, with or without IPTG, colonies will be white.
   b. Cells with the genotype $i^so^cz^+$ still have the super-repressor, but the β-galactosidase gene is under the control of a constitutive operator, o^c. This mutation interferes with the ability of the repressor to bind and repress transcription. Therefore, β-galactosidase would probably be produced continuously. On X-gal, with or without IPTG, colonies will be blue. (It is conceivable, though unlikely, that there may exist o^c mutations which do not interfere with the binding of certain super-repressors. If this were the case, the cells described here would not produce β-galactosidase, i.e., they would have the same phenotype described in part a.)
   c. Merodiploids with the genotype $i^so^cz^-/i^+o^+z^+$ would behave like the mutant described in part a above. The constitutive operator, o^c, is insensitive to the *lac* repressor, but it does not result in constitutive synthesis of β-galactosidase because it is in *cis* with an inactive β-galactosidase allele (z^-). The operator which controls a *lacZ* gene that could yield active β-galactosidase (z^+) is wild-type (o^+), and would therefore be sensitive to the super-repressor synthesized from the i^s allele. Recall that the repressor is a diffusible gene product, and can act in *trans*, unlike the operator.
   d. Merodiploid cells with the genotype $i^+o^cz^-/i^-o^+z^+$ would exhibit β-galactosidase regulation essentially identical to that of wild-type cells. Because there is only one functional copy of the repressor (i^+), and because the only functional copy of the *lacZ* gene is under the control of a wild-type operator (o^+), β-galactosidase production would be repressed in the absence of inducer when the repressor is bound to the operator. On X-gal alone, colonies would be white. In the presence of inducer, the β-galactosidase gene would be induced: on X-gal and IPTG, colonies would be blue.

5. Given the operon structure and mode of regulation described, the activities of operon products for the various mutant classes may be estimated at the following values:

	Uninduced		Induced	
Strains	*Enz A*	*Enz B*	*Enz A*	*Enz B*
Haploid strains				
(1) $R^+O^+A^+B^+$	1	1	100	100
(2) $R^+O^cA^+B^+$	1–100	1–100	100	100
(3) $R^-O^+A^+B^+$	100	100	100	100
Diploid strains				
(4) $R^+O^+A^+B^+/R^+O^+A^+B^+$	2	2	200	200
(5) $R^+O^cA^+B^+/R^+O^+A^+B^+$	2–101	2–101	200	200
(6) $R^+O^+A^-B^+/R^+O^+A^+B^+$	1	2	100	200
(7) $R^-O^+A^+B^+/R^+O^+A^+B^+$	2	2	200	200

Mutations of the O^c type, resulting in a reduced affinity of the operator region for repressor, would be expected to be *cis* dominant and characterized in both haploid and diploid strains by a constitutive expression of the *cis* operon. The uninduced level of expression would depend on the degree of residual affinity of the operator for repressor. Mutations of the R^- type, resulting in a deficiency of repressor activity, would be expected to be recessive and characterized in a haploid strain only by a constitutive expression of the operon. The uninduced level of expression in an R^- mutant would be expected to be comparable to that in the induced condition.

7. It is generally accepted that the extent of homology of a promoter to the consensus –35 and –10 sequences is an important determinant of promoter strength. It might, therefore, be reasonably expected that genes for which transcripts are needed in great amount would possess promoters with sequences very close to the consensus. Although this is in some cases true, it is not generally the case. The explanation for the less than perfect match of most promoters to the consensus sequence is to be found in the need to regulate transcription. Transcriptional regulation is achieved in many instances by the selective improvement of the affinity of specific promoters for RNA polymerase. Such selective improvement is well illustrated in the case of regulation of the *lac* operon. The *lac* promoter is by all accounts a rather weak one; its match to the consensus sequence is as follows:

	–35	–10
Promoter consensus sequence	TTGACA	TATAAT
Wild-type *lac* promoter	TTTACA	TATGTT
*lac*UV5	TTTACA	TATAAT

The strength of the wild-type *lac* promoter, and hence the level of expression of the *lac* operon, may, however, be significantly increased as a result of binding of the catabolite activator protein (CAP) upstream of the –35 region. This is because of the additional affinity of the promoter for RNA polymerase provided by the latter's interaction with CAP. A mutant containing two base replacements in the –10 region of the *lac* promoter (UV5), which dramatically improves the homology of the promoter to the consensus sequence, increases its inherent strength to such a degree that it no longer requires CAP binding for high-level transcription. As a consequence, expression of the *lac* operon in this mutant is no longer subject to regulation by catabolite repression, a situation that is clearly not advantageous to the cell when both glucose and lactose are available as carbon sources.

9. The repressor binds to its operator much better than predicted by a simple bimolecular reaction limited by diffusion in three dimensions. Thus, the lac repressor must find its operator by some other mechanism. One possibility is the binding of the repressor nonspecifically to the DNA and then searching in one dimension (binding to the DNA and then sliding along until the promoter is reached). Also, a long section of DNA would not be randomly distributed but would form a loose ball of DNA that would define a domain much smaller than the solution in the test tube. When the repressor was released from the DNA it could more quickly find another strand of DNA to bind (effectively giving a much higher concentration of DNA). The effect of both mechanisms would be to allow as few as 10 repressor molecules per cell to prevent transcription of the *lac* operon.

11. The synthesis of ribosomal proteins is regulated in *E. coli* by translational regulation, i.e., free ribosomal proteins inhibit the translation of their own mRNA. As long as rRNA is being made,

these proteins bind to the rRNA and the translation of the ribosomal proteins continues. The genes for the ribosomal proteins are clustered in a number of operons that produce polycistronic mRNAs. One of the simplest operons is P_{L11} which codes for proteins L1 and L11. L1 is the regulatory protein and can bind to the 23S rRNA or the 5' end of its own polycistronic mRNA. If the levels of L1 increase, it binds to its own mRNA and inhibits translation of both L1 and L11 proteins. This mechanism keeps the levels of L1 and L11 in register with the amount of rRNA. The other ribosomal proteins are regulated in a similar manner.

13. The process of translation is itself an amplification process, in that each molecule of ribosomal protein mRNA can give rise to many copies of the corresponding protein.
15. The genes encoded by bacteriophage lambda DNA are organized such that gene products which are required for the same function or process are clustered. This clustering of genes facilitates regulation because all of the gene products required at the same time can be simultaneously induced. The clustering also leads to more efficient organization and tighter packing of the bacteriophage genome. In other words, if the lambda genome were not so well organized, each gene would have to have its own control regions and it is likely that additional regulatory proteins would be required. As a result, the genome size would be considerably larger than it is now.
17. The two-site model for binding of regulatory proteins was suggested by the DNA binding site having a twofold axis of symmetry, which matches the symmetry of the regulatory protein. This symmetry matching is a recurring theme in DNA-protein interactions. The final proof of the two-site model came with the cocrystallization of the regulatory protein and its DNA-binding site. The protein binds on one side of the DNA in two adjacent major grooves. The hydrogen-bonding groups exposed in the major groove present many possibilities for interactions with the amino acid side chains of the protein (see fig. 30.28 in the text for the repressor binding). A common element of these proteins that bind to specific sites in DNA is the *helix-turn-helix* motif.

31 Regulation of Gene Expression in Eukaryotes

Summary

Some mechanisms of gene expression are found in eukaryotes but are rarely, if ever, seen in prokaryotes. Still, *trans*-acting regulatory proteins that bind to *cis* effector sites on the genome are present in eukaryotic systems as well as in *E. coli.* The best understood unicellular eukaryote is the budding yeast *Saccharomyces cerevisiae.* Gene regulation, particularly of development, can be quite complex in multicellular eukaryotes. Our discussion in this chapter focused on the following points.

1. In eukaryotes, individual genes encode transcripts for a single polypeptide chain. Although functionally related genes are clustered, each gene has its own promoter.
2. In yeast, genes of the *GAL* system are under the joint control of two *trans*-acting genes that encode regulatory proteins: *GAL4* and *GAL80.* GAL4 protein is an activator that binds at an upstream site, and GAL80 is a repressor that inhibits GAL4 action by binding to it. The ability of the GAL80 protein to bind to GAL4 is lost in the presence of galactose, which binds to the GAL80 protein, causing a reversible allosteric change in its structure.
3. Yeast has two haploid cells of opposite mating types and one nonmating diploid cell that results from the fusion of haploid cells of opposite mating type. The mating type is determined by the *MAT* locus. Information for the mating type is stored at other, silent loci and is expressed only if it is transposed to the *MAT* locus. Mating-type information is not expressed at the storage loci because of a complex repressor system of proteins interacting in *cis* fashion over a considerable distance at the control centers of the storage loci.
4. The greatest difference between the regulatory systems in yeast and *E. coli* is that yeast regulatory proteins can bind at a long distance from the RNA polymerase-binding site and still be effective.
5. Complex multicellular eukaryotes differentiate irreversibly so that different cell types express a different profile of genes. Genes that are expressed are usually associated with swollen chromatin. Proteins found in active regions of the genome show characteristic modifications.
6. Enhancers are elements of the genome that generally stimulate transcription. They resemble yeast UAS sequences in that they can function over long distances. In fact, they can function over even greater distances than UAS sequences, and they are effective in either orientation: Either upstream or downstream from the promoter. A possible mechanism for enhancer and UAS function over long distances is that the chromosome folds to bring the proteins bound at the enhancer site in close proximity to other regulatory proteins or RNA polymerase bound at the promoter.
7. A typical enhancer is composed of a cluster (usually two or three) of *cis* sites called enhansons; each enhanson contains binding sites for a unique combination of regulatory proteins. Enhansons must be clustered to be effective.
8. In chapter 30 we discussed DNA-binding proteins that regulate transcription in prokaryotes. In prokaryotes and eukaryotes specific recognition is dominated by H bond interactions that take place in the major groove of the DNA. In both cases the α helix is the most common element used for DNA recognition. The most striking difference between DNA-binding proteins in prokaryotes and eukaryotes has to do with the symmetry of the interaction. In prokaryotes the binding proteins almost always interact in a symmetrical fashion with the DNA. In eukaryotes most of the cases that have been examined so far involve proteins that interact in an asymmetrical fashion with the DNA.

In many cases the regulatory proteins interact in multisubunit complexes that contain nonidentical subunits. Four different types of structural motifs are discussed: The homeodomain, the zinc finger, the leucine zipper, and the helix-loop-helix.

9. Posttranscriptional regulation is an important mode of regulation in eukaryotes. Examples are given of three types of posttranscriptional regulation: Alternative modes of mRNA splicing, regulation at the translation level, and alternative modes of polypeptide processing.
10. Special combinations of regulatory factors give rise to developmental patterns of cell differentiation. Some organisms must amplify genes to keep pace with the high demand for certain gene products, especially in early development. Histones, ribosomal proteins, and ribosomal RNA genes are amplified.
11. The existence of many kinds of regulatory mutants has helped to advance our understanding of early development in the fruit fly, *Drosophila melanogaster*. Regulatory gene products are proteins that activate or repress other genes. Early development in *Drosophila* is a sequence of events in which different regulatory proteins gradually come into play in cascade fashion, controlling a wide range of enzymes and structural proteins and also influencing each other. Until the blastoderm stage the nuclei in a developing *Drosophila* embryo are not separated by cellular membranes. As a result, the regulatory proteins and other gene products may diffuse freely from their site of synthesis to other nuclei in the embryo. At the late blastoderm stage, the nuclei become cellularized. From this point on, the influence of regulatory proteins made in one cell must be exerted on another cell at the level of the cell membrane.

Problems

1. On the basis of what you know about genes in prokaryotic and eukaryotic cells, define a gene. Make sure that your definition is brief and concise. Does your definition have any limitations or problems?

2. Why is attenuation control in eukaryotes unlikely?

3. The domain-swap experiment illustrated in figure 31.4 demonstrated that the *lexA* DNA-binding domain from *E. coli* is functionally interchangeable with the equivalent domain from the GAL4 protein if and only if the *lexA* DNA recognition site replaces the GAL4 equivalent. Is the *GAL1* gene with a *lexA* binding site upstream transcribed in the presence of intact lexA protein (with no GAL4 activation domain)? Why or why not?

4. A *GAL4* mutation (*GAL4*[c]) leads to constitutive synthesis of the *GAL1* gene product in haploid yeast. Propose an explanation for the effect of this mutation.

5. How do you expect a deletion of *HMLE* to affect the expression of mating-type genes in yeast? Compare this effect with the deletion of the α_2 gene from MAT_α. Consider both homothallic and heterothallic backgrounds.

6. Color blindness is X-chromosome linked. Bearing in mind the phenomenon of X-chromosome inactivation, suggest an explanation for the observation that females who are heterozygous for the defective gene show no signs of color blindness.

7. Briggs and King were able to grow a differentiating embryo from an egg of *Rana pipiens,* the chromosomes of which were replaced with a single diploid nucleus from another embryo. What does this tell us about the state of the nucleus in the developing embryo?

8. There is no large difference in the frequency of histones in transcribed regions of the genome compared with untranscribed regions. Why don't nucleosomes interfere with transcription in eukaryotes?

9. High salt concentrations weaken the interaction of histones with DNA but have little effect on the binding of many regulatory proteins. Explain this observation in terms of how these molecules interact with DNA.

10. List some characteristics that distinguish active from inactive chromatin.

11. The restriction endonuclease *Hpa*II cleaves the sequence CCGG only if the second C is unmethylated. The enzyme *Msp*I cleaves the same sequence, whether or not it is methylated. How do the globin-specific sequences in erythroblast DNA (erythroblasts are red blood cell precursors) differ from those of other tissues in their susceptibility to these two enzymes?

12. Explain how a DNA sequence (enhancer sequence) located 5,000 bp from a gene transcription start site can stimulate transcription even if its orientation is reversed.

13. Discuss the types of structural motifs found in eukaryotic transcription factors.

14. What kind of changes have to be made in a typical eukaryotic structural gene for its protein product to be expressed in bacteria?

15. In the *Xenopus* oocyte a large number of ribosomes are made in a short time to handle the rapid demand for cell growth during cleavage stages. How is this large amount of rRNA made in such a short time?

16. Based only on the definition of maternal-effect genes, segmentation genes, and homeotic genes, which do you predict act earliest in development of the *Drosophila* embryo, and which act latest?

17. Given that specific subsets of homeotic genes are required for the development of specific segments of the *Drosophila* embryo, suggest a possible mechanism whereby the necessary spatially restricted pattern of homeotic gene expression might be achieved. Incorporate the observed effect of homeotic mutations on segment morphology into your model.

18. Given a cloned fragment of a *Drosophila* gene, how could you determine which chromosomal band(s) contain the gene?

Solutions

1. We use the term "gene" often in biology and biochemistry and take for granted that we understand what a gene is. Yet, if you asked 50 scientists their definition of a gene you would probably get 50 different answers. Our concept of a gene has evolved dramatically from the original concepts proposed by Mendel, especially with the modern approaches used in molecular biology. Our definition is: *the DNA (or RNA) sequences necessary to produce a peptide (or RNA).* Some viruses have RNA as their genetic material and some genes do not code for a protein but make a functional RNA such as tRNA or rRNA. This definition may be too general. Would it include enhancer sequences necessary for transcription located thousands of base pairs upstream or downstream from the gene? Does this definition include introns and all the regulatory sites located in the promoter region? How would polyproteins be viewed (a single peptide that is processed into a number of peptide products)? How would you explain genes produced from the same DNA region that, because of two different reading frames, generate two different polypeptides? How about different types of DNA structure such as DNA bending, Z-DNA, and higher-order chromatin structure in eukaryotes? Obviously, a single final definition of a gene remains uncertain, but the definition given covers most of our current concepts of a gene.
3. No, this construct would not be activated by binding of lexA because lexA does not have a eukaryotic activation domain. Although lexA is a repressor of bacterial transcription, even a bacterial transcriptional activator (like CAP) is unable to interact with the eukaryotic transcriptional apparatus, and therefore could not substitute for GAL4.

This control experiment shows that specific binding of a protein to upstream DNA is not sufficient to activate GAL1 transcription. The GAL4 activation domain must also establish another interaction, presumably with a protein component of the transcription apparatus. These results further confirm the bifunctional nature of GAL4 protein. Like most eukaryotic transcriptional activators, it has one domain whose function is to bind its DNA target. This domain can be exchanged with the equivalent domain of any other DNA-binding protein, even the bacterial lexA protein. The resulting hybrid protein will exhibit altered DNA-binding specificity, but transcriptional activation will be unaffected. If the transcriptional activation domain is removed, it can only be replaced with that of a similar eukaryotic transcriptional activator, otherwise it is inactive.

5. In wild-type haploid yeast only one of the HML_α and HMR_a sets of genes are expressed at any one time. This exclusivity of expression is conditioned by a transposition mechanism that removes either set of genes from an inactive "storage" chromosomal context (HML_α and HMR_a loci) to an active one (the *MAT* locus). The inactivity of HML_α and HMR_a genes at their "storage" loci is due to the close proximity of "silencer" elements (*HMLE* for HML_α and *HMRE* for HMR_a) that repress transcription of surrounding genes. Since the simultaneous expression of HML_α and HMR_a genes, which occurs in the diploid, prevents mating, it is essential that these "storage" loci be tightly repressed, in order to ensure that haploid cells display appropriate mating behavior. A deletion of the *HMLE* would remove the element repressing transcription of HML_α at the "storage" location, and result in the constitutive expression of HML_α genes from this locus. The consequences of such constitutivity would depend on the mating type of the mutant. If HML_α were at the *MAT* locus (MAT_α), the arrangement characteristic of the α mating type, the cell exhibits the behavior expected of the α mating type. However, if HMR_a were at the *MAT* locus (MAT_a), the arrangement giving rise to the *a* mating type, the cell would resemble the diploid, expressing both *a* and α genes simultaneously, and thus be sterile.

Mutants that began as α-type would become sterile at a rate conditioned by the frequency of transposition of the HMR_a to the *MAT* locus. In homothallic strains this frequency is very high, approximately once per cell division, while in heterothallic strains it is considerably lower.

Both HMR_a and *HML* exert their effect on mating behavior by controlling the expression of sets of genes specific to each mating type. *A*-specific genes are expressed only by *a* cells, while α-specific genes are expressed only by α cells. The way in which products of *MAT* (i.e., *HMR* at the *MAT* locus) are believed to regulate expression of *a*- and α-specific sets of genes is shown schematically as follows:

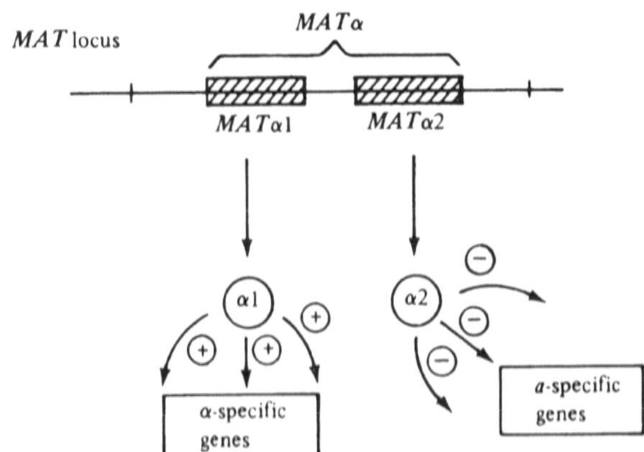

As can be seen from the figure, deletion of the gene encoding $\alpha 2$ from MAT_{α} would result in a failure to inhibit expression of the *a*-specific genes. Haploid mutants that contained such a deletion would therefore resemble the diploid, and be sterile. If the $\alpha 2$ gene contained by the HML_{α} locus were similarly mutated, mutants of this type would be unable to switch to the α-type. However, transposition of HMR_{a} to the *MAT* locus would allow for expression of the *a* mating type. Again the frequency of transposition would condition the rate of switching between sterile and *a*-type, the frequency being far greater in homothallic strains than in heterothallic strains.

7. The results of the nuclear transplantation experiments of Briggs and King indicated that nuclei, up to a certain stage in embryonic development, remain totipotent, in the sense that they can support the normal development of enucleated differentiating embryos. They found that blastula nuclei, although already "committed" to differentiated pathways, were capable of being "reprogrammed" by a proper environment to allow for some degree of dedifferentiation. Gastrula nuclei, in contrast, were found to support normal development only at a much reduced efficiency. From this they were able to conclude that these later-stage nuclei had undergone some measure of irreversible differentiation, with the result that they had lost, at least to some degree, their totipotency.
9. Histones bind to DNA by electrostatic interaction of basic amino acids (arginine and lysine) with the negative charge on the phosphate. These electrostatic (negative-positive charge) interactions are weakened by high salt, which will allow the histone proteins to disassociate from the DNA. Many regulatory proteins may initially interact in a nonspecific fashion with DNA until they locate their high-affinity binding sites. Once these proteins bind at their specific recognition sites, the major interaction is through hydrogen bonding and hydrophobic interactions. These hydrophobic interactions are stabilized by high salt.
11. MspI would cleave at every CCGG sequence in or near the globin genes, regardless of the cell type from which the DNA had been isolated, because this enzyme is insensitive to methylation. Globin genes isolated from erythroblast cells would not differ from those of other tissues in their cleavage pattern with MspI, as shown in the illustration.

    HpaII does not recognize CCGG sequences when the second C is methylated. In tissues in which genes are *not* transcribed, CpG sequences tend to be methylated more often than in tissues where the same genes are transcribed. Thus, HpaII is expected to recognize and cleave more sites in globin genes in erythroblast DNA than in DNA from other cell types (where the globin genes are not transcribed, and the corresponding DNA is more highly methylated).

    A hypothetical experiment demonstrating "hypomethylation" of actively transcribed DNA is shown below. DNA from the indicated tissues would be digested with MspI or HpaII, separated by agarose gel electrophoresis, and transferred to a membrane by the Southern blotting procedure. Hybridization of the membrane to a globin gene probe, followed by washing and autoradiography should reveal the same fragments in MspI digests of DNA from any tissue. Even in DNA isolated from erythroblasts, there may be some methylated CCGG sequences, which would therefore be resistant to HpaII cleavage (compare the MspI and HpaII lanes for erythroblast DNA). In liver DNA, considerably more CCGG sequences in and near the globin genes would be methylated, and therefore refractory to cleavage by HpaII.

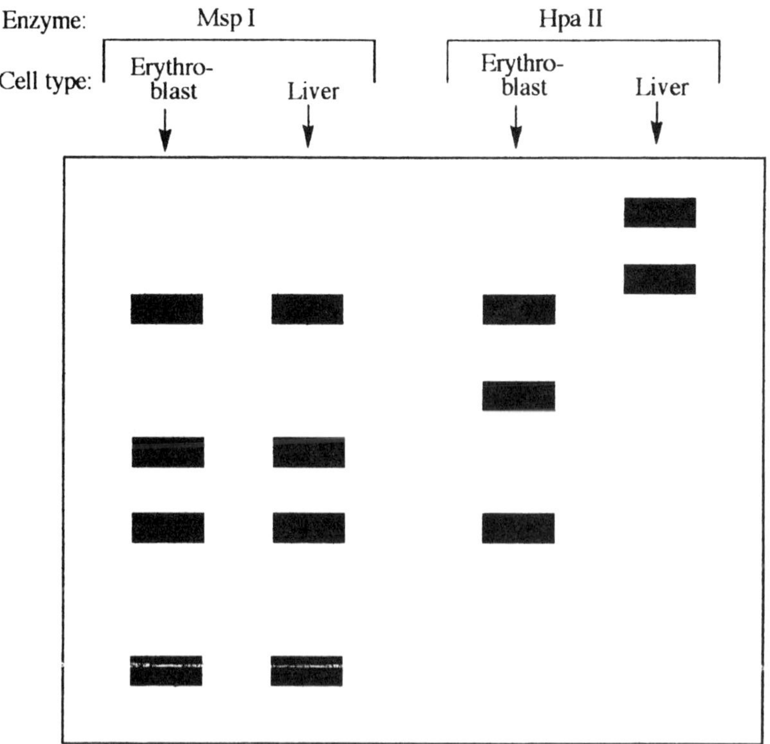

13. Most transcription factors have two-domain structures, one for DNA binding and another for protein binding. The DNA-binding domains involve various structural motifs that interact with the DNA in the major groove. The type of structure previously discussed in prokaryotes is the helix-turn-helix motif, which is also found in some eukaryotic transcription factors. A more common motif in eukaryotes is the zinc finger (zinc forms a tetrahedral complex with histidines and cysteines). Another type of structure found is the leucine zipper (a highly conserved stretch of amino acids with net basic charge followed by a region of four leucine residues at intervals of seven amino acids). All of these motifs found in these transcription factors involve interactions in the major groove of the DNA with either an α-helix segment (more common) or a two-stranded antiparallel β sheet.
15. One strategy that is employed to generate sufficient amounts of a product required at a specific stage of development is gene amplification. An example of this strategy is to be found in the sea urchin, which achieves a high rate of histone synthesis during embryogenesis by producing a large number of copies of the histone genes. Another example of gene amplification is provided by *Xenopus laevis*. The amounts of ribosomal RNA in frog eggs can be greatly increased by amplification of the rRNA genes. The amplified DNA is extrachromosomal, located near the nucleolus, transcribed during oogenesis, and subsequently discarded. The 5S rRNA genes of *Xenopus laevis* fall into two families. There are approximately 20,000 copies of a 5S RNA gene expressed during oogenesis, and about 400 copies of a 5S RNA gene expressed in somatic cells and growing oocytes. The expression of 20,000 copies of the oocyte gene provides for very high levels of 5S RNA during oocyte growth.

    An alternative strategy that is used to provide large amounts of required products early in development is the accumulation of mRNAs or proteins prior to need. This strategy is best exemplified by *X. laevis*, which stores large amounts of histone proteins and histone mRNAs in the egg to be used in embryogenesis, when the demand for histones is high.

17. Although the precise regulatory relationships between homeotic genes remain unclear, it is apparent that a hierarchy of interactions exists among gap, pair-rule, and segment polarity genes. Each gene class is influenced by the action of earlier-acting genes that control larger units of pattern, and by some members of the same class. Thus gap genes that are expressed early influence the pattern of expression of pair-rule, (e.g., fushi terazu, *ftz*), segment polarity (e.g., engrailed, *en*) and homeotic genes, (e.g., ultrabithorax, *ubx*), which are expressed later, in a hierarchical fashion. While segment polarity genes do not affect the expression of gap or pair-rule genes that occupy a higher position in the hierarchy, some gap genes, and presumably genes at other levels in the hierarchy, have been found to be mutually negative regulators of each other.

The phenomenon of mutual negative regulation offers perhaps the most plausible explanation for the observed effect of homeotic mutations on segment morphology. To date, a number of such interactions between homeotic genes have been described. The first of these involved genes of antennapedia (*Antp*) and *ubx* complexes. Mutation of the *Ubx* gene, which is expressed posterior to *Antp*, was found to result in the ectopic expression of *Antp* in those segments where *Ubx* was normally expressed, suggesting that *Ubx* represses *Antp* directly or indirectly in these segments. Subsequent studies have indicated that similar mutually negative interactions occur generally among homeotic genes.

A schematic outline of the hierarchy of interactions leading to the integrated partitioning and specification of the *Drosophila* embryo is

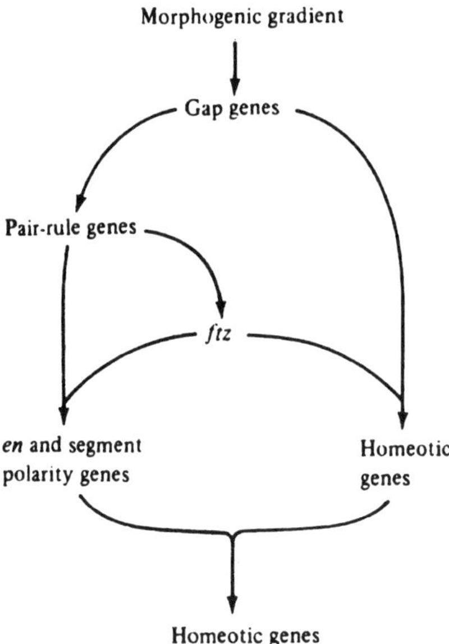

Supplement 3: Principles of Physiology and Biochemistry: Immunobiology

Summary

In this supplement we considered the major findings that led to our understanding of the immune system in vertebrates, especially mice and humans. The following points are the highlights of our discussion.

1. Two different classes of white cells (lymphocytes) are associated with the immune response: The B cells and the T cells. B cells produce antibodies that are secreted and aggregated foreign substances known as antigens. T cells are of different types: Some of them assist B cells to become antibody-forming cells; others mount an attack on antigens of their own.
2. The most common type of antibody is a tetramer composed of two identical heavy (H) chains and two identical light (L) chains. Each chain is divided into a relatively constant sequence (the *c* segments) and a relatively variable sequence (the *V* segments). Within the variable regions there are subsegments known as the hypervariable regions. These subsegments are the regions that specifically bind different antigens.
3. According to some estimates, each organism is capable of making more than a million different antibodies from a limited amount of genetic information. This feat is accomplished with the help of somatic recombination and somatic mutation. Somatic recombination involves splicing at the DNA level. It brings specific *C* and *V* regions together to make a unit that can be transcribed. Somatic mutation is focused on those parts of the gene that encode the hypervariable regions.
4. T helper cells focus the antigen on an immature B cell. In response, some B cells turn into antibody-forming cells, and some B cells turn into memory cells for production of specific antibodies at a later date. The first time an organism is exposed to a specific antigen it takes longer to mount an immunological response. That is because of the lack of memory cells for making a specific antibody.
5. Complement is a group of serum proteins that aids in the defense against microorganisms and removal of antibody-antigen complexes.
6. T cells give rise to an immune response of their own. The two types of cell-mediated immunity involve basically different types of T cells. In the delayed-type response, the T cell reacts specifically with antigens and secretes lymphokines. These are substances that attract macrophages or other leukocytes, thus producing a slowly developing inflammatory response. A second type of T cell reacts specifically with antigen bound to target cells and causes their lysis.
7. Tolerance prevents the immune system from attacking self-antigens. To understand tolerance we must appreciate how T cells work. T cells recognize a combination of self and nonself. The cell surface antigens recognized by T cells are known as the major histocompatibility complex (MHC). If two organisms carry the same histocompatibility antigens, the tissues from the two organisms are completely compatible. The histocompatibility antigens resemble antibodies in structure.

Supplement 4: Principles of Physiology and Biochemistry: Carcinogenesis and Oncogenes

Summary

From the time of their discovery, cancer cells have always appeared to be unruly. We have reached that stage in our understanding of cancer cells when we can point to specific aberrations in the genome. In this chapter we took the view that an understanding of cancer can be achieved by analyzing the regulatory pathways involved in growth control because most cancers appear to originate from mutations in specific genes involved in growth regulation. The following points are the highlights of our discussion.

1. Many properties of transformed cells grown in the tissue culture resemble cancer cells. The factors that lead to uncontrolled growth *in vivo* can be studied *in vitro* by the effects they have on tissue culture cells.
2. To judge by the frequency of occurrence of cancers in different countries, environmental factors have more influence on the incidence of cancers than inherited genetic factors do. This conclusion is reinforced by studies of migrant populations.
3. Chromosomal translocations are frequently associated with specific types of cancer. This is direct evidence that genetic abnormalities can lead to cancer.
4. A number of tumors arise from recessive mutations in which the mutations appear to be in growth-control genes.
5. Growth-control genes that lead to cancers when they are altered in some way are referred to as protooncogenes. They become oncogenes, that is, cancer-causing genes, by mutation. Host protooncogenes are frequently very similar in structure to oncogenes carried by tumor-causing viruses.
6. Dulbecco proposed that cancer-causing viruses insert their oncogenes into the host genome. It appears that cancer-causing viruses are associated with a limited number of DNA viruses and RNA viruses known as retroviruses, which replicate through a DNA intermediate. It is this DNA intermediate that usually gets inserted into the host genome.
7. Abnormal expression accompanies the transition from protooncogene to oncogene. Three types of abnormalities occur: (1) Excessive production of the gene product; (2) altered behavior of the gene product, such as a change in its regulatory properties; and (3) expression at a time during the cell cycle when the gene is not normally expressed.
8. A fully developed cancer appears to arise in steps, each step showing a breakdown in normal regulation.